PENETRANT TESTING PRINCIPLES, TECHNIQUES, APPLICATIONS, AND INTERVIEW Q&A

CHETAN SINGH

Made with ♥ on the Notion Press Platform
www.notionpress.com

Contents

Acknowledgements

The creation of this book would not have been possible without the contributions and support of many individuals and organizations.

First and foremost, we would like to thank the experts who participated in the interview Q&A sessions and generously shared their knowledge and experiences on penetrant testing. Their insights and expertise have enriched this book and provided valuable guidance to readers.

We also want to express our gratitude to the technical and editorial teams who have worked tirelessly to bring this book to fruition. Their professionalism, expertise, and dedication have ensured the quality and accuracy of the content.

Finally, we would like to thank our families and loved ones for their unwavering support and encouragement throughout the project. Their patience, understanding, and love have been indispensable in making this book a reality.

Thank you all for your contributions and support.

CHAPTER I

Chapter 1: Introduction to Penetrant Testing

Penetrant testing, also known as dye penetrant inspection or liquid penetrant testing, is a widely used non-destructive testing (NDT) technique that is used to detect surface-breaking defects on a variety of materials. The technique involves the application of a penetrating liquid, which is drawn into any surface-breaking defects, and then a developer is applied to the surface, which brings the penetrant to the surface and makes any defects visible.

Penetrant testing is a relatively simple and low-cost NDT technique, making it popular for use in many industries. It is particularly useful for detecting defects that are too small or too shallow to be detected by visual inspection and can be used on a wide range of materials, including metals, plastics, ceramics, and composites.

However, penetrant testing has its limitations and may not be effective for detecting defects that are below the surface or within porous materials. It also requires a clean surface and specific environmental conditions, such as suitable temperatures and lighting, to work effectively.

Despite these limitations, penetrant testing remains an important NDT technique that is widely used in many industries, including aerospace, automotive, and manufacturing and is a crucial tool for ensuring the safety and reliability of critical components. In the following chapters, we will explore the various aspects of penetrant testing, including the testing process, materials used, evaluation of indications, applications, and future developments.

Overview of Non-Destructive Testing (NDT)

Non-destructive testing (NDT) is a group of techniques that are used to detect defects or anomalies in materials and components without causing damage to them. These techniques are particularly useful for detecting defects that are not visible to the naked eye and can be used to evaluate the integrity and safety of critical components in a wide range of industries.

NDT techniques are used in many industries, including aerospace, automotive, construction, manufacturing, and oil and gas. They are often used to test a variety of materials, including metals, plastics, ceramics, and composites.

Some common NDT techniques include:

- Ultrasonic testing: uses high-frequency sound waves to detect flaws or defects in materials
- Radiographic testing: uses X-rays or gamma rays to detect internal flaws in materials
- Magnetic particle testing: uses magnetic fields and magnetic particles to detect surface and near-surface defects
- Liquid penetrant testing: uses a penetrating liquid to detect surface-breaking defects.

NDT techniques are often used in conjunction with each other to provide a more complete picture of the condition of a component or material. NDT is a critical tool for ensuring the safety and reliability of critical components and is often used in quality control and assurance programs.

In the following chapters, we will focus on one specific NDT technique, liquid penetrant testing, and explore its various aspects, including the testing process, materials used, evaluation of indications, applications, and future developments.

Advantages and limitations of penetrant testing

Penetrant testing, also known as dye penetrant inspection, is a non-destructive testing method used to detect surface-breaking defects in materials. This method involves applying a penetrating liquid or dye to the surface of the material being tested, allowing it to seep into any surface cracks or defects, and then applying a developer to make the dye visible. Here are some of the advantages and limitations of penetrant testing:

Advantages:

- Can detect very small defects: Penetrant testing can detect very small defects that are not visible to the naked eye. This makes it a useful tool for detecting surface cracks, porosity, and other surface defects in materials.

- Can be used on a variety of materials: Penetrant testing can be used on a variety of materials, including metals, plastics, ceramics, and composites.
- Quick and easy to perform: Penetrant testing is a quick and easy non-destructive testing method that does not require expensive equipment or specialized training.
- Low cost: Penetrant testing is a low-cost testing method, making it a practical option for many applications.

Limitations:

- Limited to surface defects: Penetrant testing can only detect surface-breaking defects, and cannot detect defects that are located beneath the surface of the material being tested.
- Cleaning is critical: The surface of the material being tested must be thoroughly cleaned before applying the penetrant, as any residual contaminants can affect the accuracy of the test.
- Requires a highly visible background: The developer used in penetrant testing requires a highly visible background to make the dye visible. This can be a limitation in cases where a highly visible background is not available or practical.
- Not suitable for rough or porous surfaces: Penetrant testing is not suitable for rough or porous surfaces, as the dye may be absorbed into the surface and make the detection of defects difficult or impossible.

Overall, penetrant testing is a valuable non-destructive testing method for detecting surface-breaking defects in a wide range of materials. However, it does have some limitations and should be used in conjunction with other testing methods to provide a more complete picture of the quality and integrity of the material being tested.

Historical background and evolution of the technique

The history of penetrant testing (PT) dates back to the early 19th century when the concept of capillary action was first discovered. The basic principle behind penetrant testing is based on this phenomenon. In 1895, the first use of penetrant testing was reported when Charles Vernon Boys used a solution of fluorescent quinine to detect surface cracks in the glass.

However, it wasn't until the 1920s that penetrant testing began to be used on a larger scale in the aerospace industry.

During World War II, the use of penetrant testing became more widespread, particularly in the aviation industry. The technique was used to detect surface cracks in aircraft components such as engine cylinders, wing spars, and landing gear. With the development of new materials and manufacturing techniques, the use of penetrant testing continued to expand into other industries such as automotive, marine, and construction.

Over time, the technique evolved and new types of penetrants were developed, such as fluorescent and visible dye penetrants, to improve the sensitivity and accuracy of the test. The development of new developer materials and application techniques also improved the visibility and ease of interpretation of the test results.

Today, penetrant testing is a widely used non-destructive testing method for detecting surface defects in a wide range of materials. It is used in a variety of industries, including aerospace, automotive, marine, and construction, to ensure the quality and integrity of critical components. The evolution of penetrant testing over time has led to improvements in the accuracy, sensitivity, and ease of use of the technique, making it an essential tool for ensuring the safety and reliability of many modern technologies.

CHAPTER II

Chapter 2: The Penetrant Testing Process

The penetrant testing process, also known as dye penetrant inspection, is a non-destructive testing method used to detect surface-breaking defects in materials. Here are the general steps involved in the penetrant testing process:

- Pre-cleaning: The surface of the material being tested must be thoroughly cleaned to remove any contaminants such as oil, grease, or dirt. This is critical to ensure that the penetrant can effectively seep into any surface cracks or defects.
- Penetrant application: A penetrant liquid or dye is then applied to the surface of the material being tested. The penetrant is allowed to sit on the surface for a specified period of time, usually between 5 and 30 minutes, to allow it to seep into any surface cracks or defects.
- Excess penetrant removal: After the specified soaking time, the excess penetrant is removed from the surface using a solvent or emulsifier. This step is important to ensure that only the penetrant that has seeped into surface cracks or defects remains on the surface.
- Developer application: A developer is then applied to the surface of the material being tested. The developer draws the penetrant out of any surface cracks or defects and makes it visible on the surface. The developer is allowed to sit on the surface for a specified period of time, usually between 10 and 30 minutes.
- Inspection: After the specified developer soak time, the surface of the material is inspected for any visible indications of surface-breaking defects. The inspector may use various lighting techniques such as UV light or white light to enhance the visibility of any defects.
- Post-cleaning: Once the inspection is complete, the developer and any remaining penetrant are removed from the surface of the material being tested using a solvent or emulsifier.

The penetrant testing process can be performed using various types of penetrants, developers, and application techniques, depending on the specific requirements of the application. The process can be used to detect surface-breaking defects in a wide range of materials, including metals, plastics, ceramics, and composites.

Pre-cleaning of the test surface

Pre-cleaning of the test surface is an important step in the penetrant testing process. The surface of the material being tested must be thoroughly cleaned to remove any contaminants such as oil, grease, or dirt before applying the penetrant. This is important for several reasons:

- Contaminants on the surface can interfere with the penetration of the penetrant into any surface cracks or defects, reducing the sensitivity and accuracy of the test.
- Contaminants can also interfere with the visibility of any defects by creating false indications or masking real defects.
- Pre-cleaning the surface ensures that any indications of surface defects are actually surface defects and not simply contaminants on the surface.
- There are several methods for pre-cleaning the surface, including solvent cleaning, emulsion cleaning, and vapor degreasing. The method chosen will depend on the type of material being tested, the contaminants present, and the specific requirements of the application.
- Solvent cleaning involves using a solvent such as acetone, isopropyl alcohol, or a proprietary solvent to dissolve and remove any contaminants on the surface. Emulsion cleaning involves using an emulsifier to break down and remove contaminants from the surface. Vapor degreasing involves exposing the surface to a vaporized solvent that condenses on the surface and dissolves any contaminants.

Regardless of the method used, it is important to ensure that the surface is completely clean and free of any contaminants before proceeding with the penetrant testing process.

Penetrant application and dwell time

Penetrant application and dwell time are important aspects of the penetrant testing process. Here's an overview of what these terms mean:

- Penetrant application: Penetrant application involves applying the penetrant liquid or dye to the surface of the material being tested. This can be done by spray, brush, or immersion, depending on the specific requirements of the application.
- Dwell time: Dwell time refers to the amount of time the penetrant is allowed to sit on the surface of the material being tested. During this time, the penetrant seeps into any surface cracks or defects, making them visible later in the testing process. The dwell time can vary depending on the type of penetrant being used, the material being tested, and the specific requirements of the application. Typically, dwell times can range from a few minutes up to 30 minutes.
- It's important to follow the recommended dwell time for the penetrant being used, as too short a dwell time may result in missed defects, while too long a dwell time can cause the penetrant to dry out and reduce the sensitivity of the test. The recommended dwell time should be specified by the manufacturer of the penetrant or in the testing standard being followed.
- After the penetrant has been applied and allowed to dwell for the recommended time, it is then removed from the surface using a solvent or emulsifier. The excess penetrant is removed to ensure that only the penetrant that has seeped into surface cracks or defects remains on the surface for later inspection.

Overall, the penetrant application and dwell time are critical steps in the penetrant testing process, as they determine the sensitivity and accuracy of the test.

Removal of excess penetrant and cleaning of the surface

After the penetrant has been allowed to dwell on the surface for a specified period of time, the excess penetrant must be removed from the surface. This is important to ensure that only the penetrant that has seeped into surface cracks or defects remains on the surface for later inspection. Here's an overview of the excess penetrant removal and surface cleaning steps in the penetrant testing process:

- Removal of excess penetrant: The excess penetrant can be removed from the surface using a solvent or emulsifier. The type of solvent or emulsifier used will depend on the specific requirements of the application and the type of penetrant being used. The excess penetrant should be removed thoroughly to ensure that only the penetrant that has seeped into surface cracks or defects remains on the surface.

- Drying time: After the excess penetrant has been removed, the surface should be allowed to dry completely. The drying time can vary depending on the specific application and the type of penetrant being used.
- Developer application: Once the surface is dry, a developer is applied to the surface. The developer draws the penetrant out of any surface cracks or defects and makes it visible on the surface. The developer is allowed to sit on the surface for a specified period of time, usually between 10 and 30 minutes.
- Inspection: After the specified developer soak time, the surface of the material is inspected for any visible indications of surface-breaking defects. The inspector may use various lighting techniques such as UV light or white light to enhance the visibility of any defects.
- Post-cleaning: Once the inspection is complete, the developer and any remaining penetrant must be removed from the surface of the material being tested using a solvent or emulsifier. This step is important to prevent any residual penetrant or developer from interfering with subsequent processes or applications.

Overall, the removal of excess penetrant and cleaning of the surface are critical steps in the penetrant testing process, as they help ensure that the test results are accurate and reliable.

Application of developer and inspection

After the excess penetrant has been removed from the surface of the material being tested, the developer is applied to the surface. The developer draws the penetrant out of any surface cracks or defects and makes it visible on the surface. Here's an overview of the developer application and inspection steps in the penetrant testing process:

- Developer application: The developer is applied to the surface of the material being tested by spray, brush, or immersion, depending on the specific requirements of the application. The developer is typically a white powder or a suspension of particles in a liquid carrier.

- Developer soak time: Once the developer has been applied to the surface, it is allowed to sit for a specified period of time, usually between 10 and 30 minutes. During this time, the developer draws the penetrant out of any surface cracks or defects, making it visible on the surface.
- Inspection: After the specified developer soak time, the surface of the material is inspected for any visible indications of surface-breaking defects. The inspector may use various lighting techniques such as UV light or white light to enhance the visibility of any defects.
- Interpretation of results: The inspector evaluates any indications of defects on the surface and determines whether they are actual surface defects or false indications caused by surface contaminants or other factors. The inspector also evaluates the size, shape, and location of any indications to determine their significance.
- Reporting: Once the inspection is complete, the results are documented and reported according to the specific requirements of the application. This may involve creating a report with details on the testing process, results, and any necessary follow-up actions.

Overall, the application of developer and inspection steps are critical in the penetrant testing process, as they help identify any surface-breaking defects and ensure the integrity and safety of the material being tested.

Interpretation of indications and reporting

Interpretation of indications and reporting are important steps in the penetrant testing process. After the surface of the material being tested has been inspected and any indications of surface-breaking defects have been identified, the inspector must interpret the indications to determine their significance. Here's an overview of the interpretation of indications and reporting steps in the penetrant testing process:

- Indication evaluation: The inspector evaluates any indications of defects on the surface to determine their significance. The inspector may use various criteria such as size, shape, location, and other factors to determine whether the indications are actual surface defects or false indications caused by surface contaminants or other factors.

- Indication classification: If the indications are determined to be actual surface defects, they are classified based on their type, size, shape, location, and other characteristics. The classification system used will depend on the specific requirements of the application.
- Indication disposition: Based on the classification of the indications, the inspector determines the disposition of the material being tested. This may involve accepting the material as-is, repairing any identified defects, or rejecting the material if the defects are deemed to be critical or cannot be repaired.
- Reporting: Once the interpretation of indications is complete, the results are documented and reported according to the specific requirements of the application. This may involve creating a report with details on the testing process, results, and any necessary follow-up actions.

Overall, the interpretation of indications and reporting steps are critical in the penetrant testing process, as they help ensure the integrity and safety of the material being tested and provide important information for follow-up actions. Proper documentation and reporting of test results can also be used to track trends, identify potential issues, and improve the overall effectiveness of the testing process.

CHAPTER III

Chapter 3: Penetrant Testing Materials

Penetrant testing requires a variety of materials to effectively identify surface-breaking defects in materials. The materials used in penetrant testing include penetrants, cleaners, developers, and auxiliary materials. Here's an overview of each:

- Penetrants: Penetrants are liquids that are applied to the surface of the material being tested and are drawn into any surface-breaking defects through capillary action. Penetrants may be classified as either visible or fluorescent, depending on the type of dye used in the formulation. Visible penetrants contain a dye that is visible to the naked eye, while fluorescent penetrants require the use of a black light to be visible.
- Cleaners: Cleaners are used to remove any surface contaminants or oils from the material being tested, ensuring that the penetrant can enter any surface-breaking defects. The type of cleaner used will depend on the specific requirements of the application and the material being tested.
- Developers: Developers are used to draw the penetrant out of any surface-breaking defects and making it visible on the surface. Developers may be classified as either dry or wet, depending on the type of formulation used. Dry developers are typically white powders that are applied to the surface and absorb any penetrant that has been drawn out of defects, while wet developers are suspensions of particles in a liquid carrier that are sprayed or brushed onto the surface.
- Auxiliary materials: Auxiliary materials may include materials such as masking tape, inspection booths, UV lights, and other items that are necessary to effectively perform penetrant testing.

Overall, the selection of penetrant testing materials will depend on the specific requirements of the application and the material being tested. Proper selection and use of these materials are critical in ensuring the accuracy and effectiveness of the testing process.

Penetrants: types, properties, and selection criteria

Penetrants are a crucial part of the penetrant testing process, as they are used to identify surface-breaking defects in materials. There are two main types of penetrants: visible penetrants and fluorescent penetrants. Here's an overview of each type, their properties, and selection criteria:

1. Visible Penetrants: Visible penetrants contain a dye that is visible to the naked eye and is used to identify surface-breaking defects. Some of the properties of visible penetrants include:

- Low sensitivity: Visible penetrants are generally less sensitive than fluorescent penetrants, making them suitable for detecting larger surface defects.
- Low cost: Visible penetrants are generally less expensive than fluorescent penetrants.
- Limited visibility: Visible penetrants may not be visible in certain lighting conditions, and the dye may be difficult to see on certain materials.

1. Fluorescent Penetrants: Fluorescent penetrants contain a dye that is not visible to the naked eye but fluoresces under UV light, making it visible on the surface of the material being tested. Some of the properties of fluorescent penetrants include:

- High sensitivity: Fluorescent penetrants are generally more sensitive than visible penetrants, making them suitable for detecting smaller surface defects.
- High visibility: Fluorescent penetrants are highly visible under UV light, making them effective for use in low-light conditions.
- Higher cost: Fluorescent penetrants are generally more expensive than visible penetrants.

When selecting a penetrant, there are several criteria to consider, including the type of material being tested, the type and size of defects being targeted, and the specific requirements of the application. Other factors, such as the required sensitivity, visibility, and cost, should also be considered when selecting a penetrant.

Overall, the selection of a penetrant is critical in ensuring the accuracy and effectiveness of the penetrant testing process, and careful consideration should be given to the specific requirements of the application and the material being tested.

Developers: types, properties, and selection criteria

Developers are a crucial part of the penetrant testing process, as they are used to draw the penetrant out of surface-breaking defects and make it visible on the surface. There are two main types of developers: dry and wet. Here's an overview of each type, their properties, and selection criteria:

1. Dry Developers: Dry developers are typically white powders that are applied to the surface of the material being tested. Some of the properties of dry developers include:

- Easy to use: Dry developers are easy to apply and do not require any special equipment.
- No drying time: Dry developers do not require any drying time, which can save time during the testing process.
- Limited sensitivity: Dry developers are generally less sensitive than wet developers, making them suitable for detecting larger defects.
- Wet Developers: Wet developers are suspensions of particles in a liquid carrier that are sprayed or brushed onto the surface of the material being tested. Some of the properties of wet developers include:
- High sensitivity: Wet developers are generally more sensitive than dry developers, making them suitable for detecting smaller defects.
- Longer drying time: Wet developers require a longer drying time than dry developers, which can increase the testing time.
- More complex to use: Wet developers require special equipment for application and can be more difficult to use than dry developers.
- In addition to the types of developers, there are also several properties and selection criteria that should be considered when choosing a developer for a specific application. Some of these include:
- Particle size: The particle size of the developer can affect its ability to draw penetrant out of defects. Finer particles are generally better at drawing out penetrants and can provide higher sensitivity.
- Color contrast: The color of the developer should provide good contrast with the color of the penetrant for effective detection of defects.

- Adhesion: The developer should adhere well to the surface being tested and should not flake or rub off during the testing process.
- Environmental factors: The selection of a developer should also take into account any environmental factors that may affect the testing process, such as temperature, humidity, and lighting conditions.
- Safety considerations: Developers can contain potentially hazardous materials, such as silica, so safety considerations should be taken into account when selecting a developer.
- Regulatory requirements: Regulatory requirements, such as those set forth by the American Society for Testing and Materials (ASTM), may specify certain requirements for the selection of developers, such as particle size or composition.

When selecting a developer, there are several criteria to consider, including the type and size of defects being targeted, the specific requirements of the application, and the type of penetrant being used. Other factors, such as the required sensitivity, visibility, and cost, should also be considered when selecting a developer.

Overall, the selection of a developer should be based on the specific requirements of the application, including the type and size of defects being targeted, the type of penetrant being used, and any environmental or safety considerations. By carefully considering these factors, it is possible to select a developer that will provide accurate and reliable results for a particular application.

Cleaning agents: types, properties, and selection criteria

Cleaning agents are used in the penetrant testing process to remove surface contaminants from the material being tested. There are various types of cleaning agents available, each with its own properties and selection criteria. Here's an overview of some common types of cleaning agents, their properties, and selection criteria:

1. Solvent-based cleaners: Solvent-based cleaners are typically used for removing oils, greases, and other non-polar contaminants. Some of the properties of solvent-based cleaners include:

- Effective for non-polar contaminants: Solvent-based cleaners are highly effective for removing non-polar contaminants, such as oils and greases.
- Quick drying: Solvent-based cleaners evaporate quickly, which can reduce the cleaning time.
- Flammable: Solvent-based cleaners are often flammable and can pose safety risks if not used properly.

2. Aqueous cleaners: Aqueous cleaners are water-based cleaners that are used for removing polar contaminants, such as salts and water-soluble oils. Some of the properties of aqueous cleaners include:

- Environmentally friendly: Aqueous cleaners are typically more environmentally friendly than solvent-based cleaners, as they do not contain harsh chemicals.
- Lower toxicity: Aqueous cleaners are generally less toxic than solvent-based cleaners and pose fewer safety risks.
- Longer drying time: Aqueous cleaners require a longer drying time than solvent-based cleaners, which can increase the cleaning time.

3. Emulsifiers: Emulsifiers are used for removing both polar and non-polar contaminants. Emulsifiers work by breaking down the contaminants and suspending them in a liquid for easy removal. Some of the properties of emulsifiers include:

- Effective for a wide range of contaminants: Emulsifiers are highly effective for removing both polar and non-polar contaminants, making them a versatile cleaning option.
- Longer cleaning time: Emulsifiers often require a longer cleaning time than solvent-based cleaners or aqueous cleaners.
- Residue: Emulsifiers can leave a residue on the surface being tested if not properly rinsed.

When selecting a cleaning agent, several criteria should be considered, including the type of contaminants to be removed, the surface being tested, the type of penetrant being used, and any safety or environmental concerns. Other factors, such as cleaning effectiveness, drying time, and cost, should also be taken into account when selecting a cleaning agent. By carefully considering these factors, it is possible to select a cleaning agent that will provide effective cleaning without compromising the accuracy and reliability of the penetrant testing process.

CHAPTER IV

Chapter 4: Surface Preparation and Test Conditions

Surface preparation and test conditions are critical factors that can impact the accuracy and reliability of penetrant testing results. Here are some important considerations related to surface preparation and test conditions:

- Surface cleanliness: Prior to performing penetrant testing, the surface of the material being tested must be thoroughly cleaned and free of any contaminants. This can be achieved through various cleaning methods, such as solvent cleaning, aqueous cleaning, or abrasive blasting.
- Surface roughness: The surface roughness of the material being tested can affect the ability of the penetrant to enter defects and the ability of the developer to form an indication. In general, the surface roughness of between 63 and 125 microinches is recommended for most penetrant testing applications.
- Test environment: The test environment can also affect the results of penetrant testing. Factors such as temperature, humidity, and lighting conditions can impact the sensitivity and visibility of indications. Testing should be performed in a controlled environment that meets the specifications of the relevant testing standard.
- Test surface condition: The condition of the test surface can also impact the accuracy of the results. For example, surface oxidation or rust can reduce the effectiveness of penetrant testing by hindering the ability of the penetrant to enter defects.
- Test object geometry: The geometry of the object being tested can also impact the accuracy of the results. Objects with complex geometries or deep recesses may require additional testing methods or techniques to ensure full coverage and reliable results.
- Test temperature and time: The temperature and time at which the penetrant is applied and allowed to dwell can also affect the results. For example, higher temperatures can reduce the dwell time required, while colder temperatures may require longer dwell times.

In summary, surface preparation and test conditions are critical factors that can impact the accuracy and reliability of penetrant testing results. By carefully considering these factors and following the relevant testing standard, it is possible to ensure accurate and reliable results for a wide range of applications.

Surface preparation and cleaning methods

Surface preparation and cleaning methods are critical to ensure that the surface of the material being tested is free of any contaminants that may interfere with the penetration of the penetrant into defects or the formation of indications. Here are some common surface preparation and cleaning methods used in penetrant testing:

- Solvent cleaning: Solvent cleaning involves using a solvent, such as acetone, to remove any oils, grease, or other contaminants from the surface of the material being tested. This method is typically used for non-porous materials such as metals.
- Aqueous cleaning: Aqueous cleaning involves using a water-based solution to clean the surface of the material being tested. This method is typically used for porous materials such as ceramics or composites.
- Emulsion cleaning: Emulsion cleaning involves using a mixture of water and oil-based emulsifier to remove any oils, grease, or other contaminants from the surface of the material being tested. This method is typically used for non-porous materials such as metals.
- Abrasive blasting: Abrasive blasting involves using a stream of abrasive particles, such as sand or glass beads, to remove any surface contaminants and create a rougher surface that promotes better penetration of the penetrant. This method is typically used for metals and other hard materials.
- Grinding: Grinding involves using a grinding wheel to remove any surface contaminants and create a rougher surface that promotes better penetration of the penetrant. This method is typically used for metals and other hard materials.
- Ultrasonic cleaning: Ultrasonic cleaning involves using high-frequency sound waves to agitate a cleaning solution, which can remove contaminants from the surface of the material being tested. This method is typically used for complex geometries or delicate materials.

In summary, surface preparation and cleaning methods are critical to ensure accurate and reliable results in penetrant testing. By selecting the appropriate cleaning method for the material being tested and following the relevant testing standard, it is possible to achieve accurate and reliable results for a wide range of applications.

Test conditions and factors affecting sensitivity

Test conditions and factors affecting sensitivity are important considerations in penetrant testing, as they can impact the ability of the penetrant to enter defects and the ability of the developer to form indications. Here are some common test conditions and factors affecting sensitivity in penetrant testing:

- Temperature: The temperature at which the penetrant is applied and allowed to dwell can affect the sensitivity of the test. Higher temperatures can reduce the dwell time required for the penetrant to enter defects and increase the overall sensitivity of the test.
- Dwell time: The length of time that the penetrant is allowed to dwell on the surface of the material being tested can affect the sensitivity of the test. Longer dwell times can increase the probability of the penetrant entering defects and forming indications.
- Penetrant type: The type of penetrant used can affect the sensitivity of the test. For example, fluorescent penetrants are generally more sensitive than visible penetrants.
- Penetrant concentration: The concentration of the penetrant can affect the sensitivity of the test. Higher concentrations of penetrant can improve the ability of the penetrant to enter defects and form indications.
- Test surface condition: The condition of the test surface can affect the sensitivity of the test. Surface oxidation or rust can reduce the effectiveness of penetrant testing by hindering the ability of the penetrant to enter defects.
- Lighting conditions: Proper lighting conditions are critical to ensure that indications are visible. The lighting conditions should meet the specifications of the relevant testing standard.
- Background contrast: The contrast between the penetrant and the background can affect the sensitivity of the test. For example,

fluorescent penetrants may require a darker background to ensure that indications are visible.

In summary, test conditions and factors affecting sensitivity are important considerations in penetrant testing. By carefully controlling these factors and following the relevant testing standard, it is possible to achieve accurate and reliable results for a wide range of applications.

Test specifications and standards

Test specifications and standards are important in penetrant testing to ensure that the testing is performed consistently and to a high standard. Here are some common test specifications and standards used in penetrant testing:

- ASTM E1417: Standard Practice for Liquid Penetrant Testing
- ASME BPVC Section V, Article 6: Liquid Penetrant Examination
- MIL-STD-6866: Inspection, Liquid Penetrant
- ISO 3452-1: Non-destructive testing — Penetrant testing — Part 1: General principles
- EN 571-1: Non-destructive testing — Penetrant testing — Part 1: General principles
- NAS 410: Certification and Qualification of Nondestructive Personnel

These specifications and standards provide guidelines for the selection of materials, the preparation of the surface, the application of penetrant, the use of developers, the inspection and interpretation of results, and the reporting of findings. They also provide guidance on the training and certification of personnel performing the testing.

In addition to these standards, there may be specific customer or industry requirements that must be met, such as aerospace or automotive industry standards.

It is important to follow the relevant testing standard and any additional requirements to ensure that the testing is performed consistently and to a high standard. This will help to ensure that accurate and reliable results are obtained and that any defects are detected and addressed in a timely manner.

Penetrant testing standards also provide guidance on the qualification and certification of personnel involved in the testing process. This includes the qualification of Level I, Level II, and Level III personnel who perform penetrant testing, interpret test results, and supervise the testing process. Personnel qualification typically involves a combination of classroom training, on-the-job training, and examination.

In addition to personnel qualification, standards such as ASME BPVC Section V, Article 6 and NAS 410 also provide guidelines for the periodic re-qualification and recertification of personnel. This helps to ensure that personnel remains competent and up-to-date with the latest testing procedures and techniques.

Other important considerations in penetrant testing standards include safety requirements and environmental considerations. Penetrant testing can involve the use of chemicals that can be hazardous to personnel or the environment if not handled properly. Standards such as ASTM E1417 provide guidance on safety requirements, including the use of personal protective equipment, and environmental considerations, such as the proper disposal of waste materials.

In summary, penetrant testing standards provide guidance on all aspects of the testing process, from the selection of materials to the qualification of personnel to safety and environmental considerations. By following these standards, it is possible to perform accurate and reliable penetrant testing and ensure that defects are detected and addressed in a timely manner.

CHAPTER V

Chapter 5: Evaluation of Indications

Evaluation of indications is a critical step in the penetrant testing process. The indications are the marks or indications left on the surface after the penetrant and developer have been applied and removed. These indications may be caused by discontinuities or defects in the material being tested, or they may be caused by other factors such as surface roughness or staining.

Evaluation of indications involves a careful visual inspection of the surface using appropriate lighting and magnification. The inspector must be trained to recognize and interpret the various types of indications that may be present, including cracks, porosity, inclusions, and other types of discontinuities.

There are several factors that must be considered when evaluating indications:

- Location: The location of the indication may help to determine its significance. Indications located in critical areas, such as stress concentration areas or areas of high stress, may be more significant than indications located in non-critical areas.
- Size: The size of the indication may also help to determine its significance. Larger indications may be more significant than smaller indications, particularly if they are located in critical areas.
- Orientation: The orientation of the indication may also be important. For example, indications that are oriented perpendicular to the direction of stress may be more significant than indications that are oriented parallel to the direction of stress.
- Context: The context in which the indication is found may also be important. For example, if there are multiple indications in a particular area, this may indicate a more significant problem than a single isolated indication.

Based on these factors, the inspector will determine whether the indications are acceptable or whether they represent defects that require further action, such as repair or rejection of the part.

It is important to note that the evaluation of indications is a subjective process that relies on the inspector's skill and experience. Therefore, it is important to ensure that inspectors are properly trained and qualified to perform the evaluation of indications. This typically involves a combination of classroom training, on-the-job training, and practical experience. In addition, periodic requalification and recertification may be required to ensure that inspectors remain competent and up-to-date with the latest testing procedures and techniques.

Types of indications and their causes

Indications are the marks or patterns that are formed on the surface of the part being tested during penetrant testing. These indications may be caused by a variety of factors, including surface conditions, manufacturing processes, and material properties. Some common types of indications and their causes include:

- Cracks: Cracks are one of the most common types of indications found during penetrant testing. They may be caused by a variety of factors, including stress, fatigue, and improper manufacturing processes. Cracks may be oriented in different directions and may appear as straight lines or irregular shapes.
- Porosity: Porosity is another common type of indication that may be found during penetrant testing. Porosity is caused by the presence of voids or air pockets within the material being tested. It may be caused by a variety of factors, including improper casting, welding, or other manufacturing processes.
- Inclusions: Inclusions are foreign materials that are embedded within the material being tested. They may be caused by a variety of factors, including improper manufacturing processes or contamination during production.
- Welding defects: Welding defects may include lack of fusion, incomplete penetration, or porosity within the weld. These defects may be caused by improper welding techniques or improper selection of welding materials.
- Machining defects: Machining defects may include surface scratches, tool marks, or other surface irregularities that may be mistaken for cracks or other defects during penetrant testing.

- Corrosion: Corrosion may appear as a series of small, irregularly-shaped pits on the surface of the material being tested. Corrosion may be caused by exposure to moisture, chemicals, or other corrosive substances.
- Fatigue: Fatigue cracks are caused by repeated cycles of loading and unloading of the material. These types of cracks usually occur at or near the surface of the material and may appear as fine lines or branching patterns.
- Surface discontinuities: Surface discontinuities may include scratches, gouges, nicks, or other surface irregularities that may be caused by handling or other surface damage. These types of indications may be mistaken for cracks or other defects during penetrant testing.
- Heat treatment cracks: Heat treatment cracks may appear as fine, hairline cracks on the surface of the material. They are caused by improper heat treatment processes, such as quenching or tempering.
- Cold working: Cold working may cause surface deformation, which can lead to the formation of cracks or other defects during penetrant testing. These types of indications may be mistaken for actual defects, so it is important to carefully evaluate them to determine their significance.

It is important to note that the presence of an indication does not necessarily mean that the part is defective. Indications must be carefully evaluated based on their location, size, orientation, and context in order to determine whether they represent a significant defect that requires further action.

It is important to note that the type and cause of indications may vary depending on the material being tested, the manufacturing processes used, and other factors. Therefore, it is important to carefully evaluate all indications and to use a combination of testing methods to fully understand the nature and significance of any defects that are found.

Indication classification and evaluation

Indications found during penetrant testing are typically classified based on their size, shape, and location. This classification system helps to determine the significance of the indication and whether further action is necessary. There are several commonly used classification systems, including the following:

- ASTM E1417: This classification system is widely used in the aerospace industry and divides indications into three categories: type 1, type 2, and type 3. Type 1 indications are the smallest and least significant, while type 3 indications are the largest and most significant.
- ASME BPVC Section V: This classification system is used in the pressure vessel industry and divides indications into four categories: category A, category B, category C, and category D. Category A indications are the smallest and least significant, while category D indications are the largest and most significant.
- MIL-STD-6866: This classification system is used by the military and divides indications into four categories: class 1, class 2, class 3, and class 4. Class 1 indications are the smallest and least significant, while class 4 indications are the largest and most significant.

Once indications have been classified, they must be evaluated to determine their significance. This evaluation involves considering several factors, including the size, shape, and location of the indication, as well as the material being tested and the specific application. Some indications may be acceptable within certain limits, while others may require further evaluation or corrective action.

In addition to evaluating the indications themselves, it is important to evaluate the overall performance of the penetrant testing process. This includes ensuring that the testing equipment is functioning properly, that the correct testing procedures are being followed, and that the testing personnel is properly trained and qualified. By carefully evaluating indications and monitoring the testing process, it is possible to ensure that the highest levels of quality and safety are maintained.

Verification of indications by complementary NDT techniques

In some cases, indications found during penetrant testing may need to be verified using complementary non-destructive testing (NDT) techniques. This can be particularly important for indications that are difficult to interpret or that require further evaluation.

Some of the complementary NDT techniques that may be used to verify indications found during penetrant testing include the following:

- Radiography: Radiography uses X-rays or gamma rays to produce images of the internal structure of a material. This technique can be used to verify indications found during penetrant testing by providing additional information about the size, shape, and location of the indication.
- Ultrasonic Testing: Ultrasonic testing uses high-frequency sound waves to detect internal defects in materials. This technique can be used to verify indications found during penetrant testing by providing additional information about the size, shape, and location of the indication, as well as the depth and orientation of the defect.
- Magnetic Particle Testing: Magnetic particle testing uses magnetic fields and magnetic particles to detect surface and near-surface defects in materials. This technique can be used to verify indications found during penetrant testing by providing additional information about the size, shape, and location of the indication, as well as the direction and orientation of the defect.
- Eddy Current Testing: Eddy current testing uses electromagnetic fields to detect surface and near-surface defects in conductive materials. This technique can be used to verify indications found during penetrant testing by providing additional information about the size, shape, and location of the indication, as well as the depth and orientation of the defect.

By using complementary NDT techniques to verify indications found during penetrant testing, it is possible to ensure that the highest levels of quality and safety are maintained. It is important to carefully evaluate all indications and to use the most appropriate NDT techniques for each situation to ensure accurate and reliable results.

CHAPTER VI

Chapter 6: Applications of Penetrant Testing

Penetrant testing is widely used in various industries for the detection of surface-breaking defects in a wide range of materials. Some of the common applications of penetrant testing are as follows:

- Aerospace industry: Penetrant testing is used extensively in the aerospace industry for the detection of surface defects in critical components such as engine components, landing gear, and other parts that are subject to high stress and fatigue. It is also used for the inspection of aircraft structures and surfaces.
- Automotive industry: Penetrant testing is used in the automotive industry for the detection of surface defects in components such as engine blocks, transmission components, and other critical parts. It is also used for the inspection of welds and other components during manufacturing and assembly.
- Manufacturing industry: Penetrant testing is used in the manufacturing industry for the inspection of a wide range of components and structures. It is used for the detection of surface defects in parts such as valves, pumps, and other components used in various industries.
- Power generation industry: Penetrant testing is used in the power generation industry for the inspection of components such as turbines, generators, and other critical parts. It is also used for the inspection of welds and other components during manufacturing and assembly.
- Medical industry: Penetrant testing is used in the medical industry for the inspection of components such as implants, surgical instruments, and other critical parts. It is also used for the inspection of welds and other components during manufacturing and assembly.
- Railway industry: Penetrant testing is used in the railway industry for the inspection of components such as rails, wheels, axles, and other critical parts. It is also used for the inspection of welds and other components during manufacturing and assembly.
- Oil and gas industry: Penetrant testing is used in the oil and gas industry for the inspection of components such as pipelines, tanks, and other critical parts. It is also used for the inspection of welds and other

components during manufacturing and assembly.

Overall, penetrant testing is a versatile and widely used non-destructive testing technique that is essential for the inspection and quality control of a wide range of components and structures in various industries.

Aerospace and aviation industry applications

Penetrant testing is an important non-destructive testing technique used in the aerospace and aviation industry for the detection of surface-breaking defects in critical components such as engine components, landing gear, and other parts that are subject to high stress and fatigue. It is also used for the inspection of aircraft structures and surfaces.

In the aerospace industry, penetrant testing is used for the inspection of various components such as engine components, turbine blades, landing gear, wing structures, and other critical parts. It is also used for the inspection of welds, fasteners, and other components during manufacturing and assembly.

One of the key benefits of penetrant testing in the aerospace industry is that it can detect surface-breaking defects that may not be visible to the naked eye. This is especially important for critical components that are subject to high stress and fatigue, as even small defects can lead to catastrophic failures.

Penetrant testing is also used for the inspection of aircraft structures and surfaces. This includes the inspection of wing structures, fuselage, and other components for cracks, corrosion, and other defects. Penetrant testing is also used for the inspection of coatings and surface treatments used in the aerospace industry.

Overall, penetrant testing is a critical technique in the aerospace and aviation industry for ensuring the safety and reliability of aircraft and other components. It is widely used during the manufacturing, assembly, and maintenance of aircraft to detect defects and ensure that critical components are free from defects that could compromise their performance or safety.

Automotive industry applications

Penetrant testing is also used in the automotive industry for the inspection of critical components such as engine blocks, cylinder heads, crankshafts, and other components. It is used to detect surface-breaking defects such as cracks, porosity, and other defects that can compromise the performance and safety of these components.

In the automotive industry, penetrant testing is used during the manufacturing and assembly of engine components to ensure that they are free from defects. It is also used during maintenance and repair operations to inspect engine components for defects that can occur over time.

Penetrant testing is also used for the inspection of welding and other joining processes used in the automotive industry. This includes the inspection of welds in the chassis, body, and other components to ensure that they are free from defects that can compromise the structural integrity of the vehicle.

Another application of penetrant testing in the automotive industry is the inspection of castings and forgings used in the manufacturing of critical components. This includes the inspection of engine blocks, cylinder heads, and other components for defects that can occur during the manufacturing process.

Overall, penetrant testing is a critical technique in the automotive industry for ensuring the performance and safety of critical components used in vehicles. It is widely used during the manufacturing, assembly, and maintenance of vehicles to detect defects and ensure that critical components are free from defects that could compromise their performance or safety.

Power generation industry applications

Penetrant testing is used in the power generation industry for the inspection of critical components such as turbine blades, steam pipes, heat exchangers, and other components. It is used to detect surface-breaking defects such as cracks, porosity, and other defects that can compromise the performance and safety of these components.

In the power generation industry, penetrant testing is used during the manufacturing and assembly of components to ensure that they are free from defects. It is also used during maintenance and repair operations to inspect components for defects that can occur over time.

Penetrant testing is also used for the inspection of welding and other joining processes used in the power generation industry. This includes the inspection of welds in pipelines, boilers, and other components to ensure that they are free from defects that can compromise their structural integrity.

Another application of penetrant testing in the power generation industry is the inspection of coatings and surface treatments used to protect components from corrosion and other types of degradation. Penetrant testing is used to detect defects in coatings and surface treatments, which can compromise their effectiveness in protecting components from degradation.

Overall, penetrant testing is a critical technique in the power generation industry for ensuring the performance and safety of critical components used in power plants. It is widely used during the manufacturing, assembly, and maintenance of components to detect defects and ensure that critical components are free from defects that could compromise their performance or safety.

Pipeline and pressure vessel inspections

Penetrant testing is used for the inspection of pipelines and pressure vessels in the oil and gas industry, chemical processing plants, and other industries that use high-pressure vessels and pipelines. The purpose of this inspection is to detect surface-breaking defects such as cracks, porosity, and other defects that can compromise the safety and reliability of the pipelines and pressure vessels.

In pipeline inspections, penetrant testing is used to inspect welds, flanges, and other components for defects that can occur during the manufacturing process or over time due to wear and tear. This inspection is critical for ensuring the safety and reliability of the pipelines, which transport hazardous materials over long distances.

Penetrant testing is also used for the inspection of pressure vessels, which are used in chemical processing plants and other industries that use high-pressure systems. This inspection is critical for ensuring the safety and reliability of the pressure vessels, which are subject to high stresses and strains during operation.

In both pipeline and pressure vessel inspections, penetrant testing is often used in conjunction with other non-destructive testing techniques such as radiography, ultrasonic testing, and magnetic particle testing to provide a comprehensive inspection of the components. This approach helps to ensure that all defects are detected, even those that may not be visible to the naked eye.

Overall, penetrant testing is a critical technique for the inspection of pipelines and pressure vessels in the oil and gas industry, chemical processing plants, and other industries that use high-pressure systems. It is used to detect defects that can compromise the safety and reliability of these systems and to ensure that they meet regulatory and safety standards.

CHAPTER VII

Chapter 7: Quality Assurance and Training

Quality assurance and training are essential components of a successful penetrant testing program. Here are some important considerations for both:

Quality Assurance:

- Procedure Development: The development of detailed and standardized procedures is essential to ensure consistency and accuracy in penetrant testing. These procedures should outline the steps involved in the testing process, including surface preparation, application of penetrant, developer application, inspection, and interpretation of results.
- Certification of Personnel: The personnel responsible for performing penetrant testing should be certified in accordance with industry standards such as ASNT SNT-TC-1A or ISO 9712. Certification programs provide training and evaluation of personnel to ensure they have the knowledge and skills to perform penetrant testing accurately and effectively.
- Equipment Calibration: Equipment used in penetrant testing should be calibrated regularly to ensure accuracy and consistency in the testing process. Calibration should be performed in accordance with industry standards such as ASTM E1417.
- Record Keeping: Detailed records should be kept of all penetrant testing performed, including the procedures used, equipment used, and results obtained. These records should be maintained for future reference and analysis.

Training:

- Basic Principles of Penetrant Testing: Training should cover the basic principles of penetrant testing, including the types of penetrants and developers, surface preparation, application of penetrant, developer application, inspection, and interpretation of results.
- Procedure Familiarization: Personnel should be trained in the specific procedures used in their facility for penetrant testing. This includes

understanding the requirements for surface preparation, application of penetrant and developer, inspection, and interpretation of results.

- Equipment Operation and Maintenance: Personnel should be trained in the operation and maintenance of equipment used in penetrant testing. This includes understanding the functions of the equipment, proper use, and maintenance procedures.
- Safety Considerations: Training should include safety considerations, including the hazards associated with the materials used in penetrant testing, proper handling, and disposal procedures.

By implementing a comprehensive quality assurance program and providing ongoing training to personnel, penetrant testing can be performed accurately and effectively, providing a reliable and cost-effective non-destructive testing method for detecting defects in critical components and systems.

Quality control and quality assurance procedures

Quality control and quality assurance procedures are critical for ensuring the accuracy and reliability of penetrant testing results. Here are some essential procedures for both:

Quality Control:

- Surface Preparation: The surface preparation process is essential for the proper performance of penetrant testing. A proper cleaning process must be used to remove any dirt, oil, or other contaminants from the surface.
- Penetrant Application: Penetrant application must be performed according to the established procedure to ensure complete coverage and penetration of the material.
- Developer Application: The developer should be applied evenly and allowed to dwell for the recommended time before removal.
- Inspection: Inspection should be performed under the appropriate lighting conditions to ensure maximum contrast and clarity of the indications.
- Documentation: All test results and other relevant data should be documented and maintained for future reference.

Quality Assurance:

- Procedure Development: A documented procedure should be established that outlines the steps involved in the testing process. This procedure should be based on industry standards and tailored to the specific application.
- Personnel Qualification: The personnel responsible for performing penetrant testing should be qualified and certified according to industry standards.
- Equipment Calibration: All equipment used in penetrant testing should be calibrated regularly to ensure accuracy and reliability.
- Record Keeping: All relevant data should be documented and maintained for future reference.
- Auditing: Regular audits should be conducted to ensure that the procedures and processes are being followed correctly.

By implementing these quality control and quality assurance procedures, you can ensure that your penetrant testing program produces accurate, reliable results that meet the needs of your organization.

Personnel qualification and certification requirements

Personnel qualification and certification requirements are essential for ensuring the accuracy and reliability of penetrant testing results. The following are some essential aspects of personnel qualification and certification:

- Education and Training: Personnel involved in penetrant testing must have a good understanding of the principles of the technique, including its limitations and advantages.
- Experience: It is essential to have practical experience with the application of the technique to different materials and test conditions.
- Certification: Personnel should be certified according to industry standards such as ISO 9712 or ASNT SNT-TC-1A. These certifications require a minimum level of education, training, and experience, as well as successful completion of written and practical examinations.
- Continuing Education: Continuing education is essential to keep personnel up-to-date on changes in standards, new techniques, and new

equipment.

- Performance Evaluation: Regular performance evaluations should be conducted to ensure that personnel is performing their duties correctly and meeting the required standards.

By ensuring that personnel is qualified and certified, you can ensure that your penetrant testing program produces accurate, reliable results that meet the needs of your organization.

The specific requirements for personnel qualification and certification will depend on the industry and standards used. For example, in the aerospace industry, personnel may need to meet specific requirements outlined in standards such as NAS 410 or EN 4179. In the automotive industry, personnel may need to meet requirements outlined in industry-specific standards or customer specifications.

Training and education programs

Training and education programs for penetrant testing are available from a variety of sources, including universities, technical schools, private training organizations, and industry associations. The type and length of training programs vary depending on the level of education and certification desired.

Some common types of training programs for penetrant testing include:

- Level I Training: This training is intended for personnel who will perform penetrant testing under the supervision of a Level II or Level III certified individual. Level I training typically covers basic principles and practices of penetrant testing.
- Level II Training: This training is intended for personnel who will perform and interpret penetrant testing independently. Level II training covers more advanced principles and practices of penetrant testing, including the interpretation of indications.
- Level III Training: This training is intended for personnel who will be responsible for developing and implementing penetrant testing procedures, as well as providing technical guidance to Level I and Level II personnel. Level III training covers the most advanced principles and practices of penetrant testing.

Training programs for penetrant testing typically include a combination of classroom instruction, hands-on practice, and practical examinations. Some programs may also include online training modules or computer-based training.

Industry associations such as the American Society for Non-destructive Testing (ASNT) and the European Federation for Non-Destructive Testing (EFNDT) offer certification programs for personnel who successfully complete training and meet other certification requirements. Additionally, some industries and companies may have their own specific training and certification requirements.

CHAPTER VIII

Chapter 8: Future Developments and Emerging Technologies

Penetrant testing is a well-established and widely used non-destructive testing technique, but there is ongoing research and development to improve the technique and expand its applications. Some of the areas of research and development in penetrant testing include:

- High-temperature penetrant testing: There is ongoing research to develop penetrant testing methods that can be used at higher temperatures, such as for the inspection of gas turbine components.
- Fluorescent penetrant testing: fluorescent penetrant testing is already widely used, but there is ongoing research to improve the sensitivity and reliability of this technique, particularly for the detection of very small defects.
- Digital radiography and computed tomography: While not technically penetrant testing, these imaging techniques are sometimes used in conjunction with penetrant testing for more comprehensive defect detection.
- Electrochemical methods: There is ongoing research into the use of electrochemical methods for penetrant testing, which could potentially improve the sensitivity and reliability of the technique.
- Artificial intelligence and machine learning: As with many areas of nondestructive testing, there is growing interested in the use of artificial intelligence and machine learning algorithms for defect detection and interpretation in penetrant testing.

Overall, the future of penetrant testing is likely to involve continued improvement and refinement of existing techniques, as well as the development of new methods and technologies that can provide even greater sensitivity, reliability, and efficiency for defect detection.

Current research and development activities

There are several ongoing research and development activities in the field of penetrant testing. Some of the current research areas include:

- Alternative penetrant formulations: Researchers are developing new penetrant formulations that can improve detection sensitivity, reduce waste, and minimize environmental impacts. Some of the new formulations include water-based and solvent-free penetrants.
- Smart penetrant systems: Smart penetrant systems are being developed that can detect and quantify defects in real time. These systems use sensors to detect and measure the extent of the defect and can provide instant feedback to the operator.
- Advanced imaging techniques: Researchers are investigating the use of advanced imaging techniques such as terahertz imaging and digital holography for penetrant testing. These techniques have the potential to improve the detection sensitivity and resolution of the technique.
- Robotics and automation: Robotic systems are being developed to automate the penetrant testing process. These systems can improve the accuracy and repeatability of the testing process and reduce the risk of operator error.
- Standards development: There is ongoing work to develop and update industry standards for penetrant testing. These standards help ensure that penetrant testing is performed consistently and reliably across different industries and applications.

Overall, the current research and development activities in penetrant testing aim to improve the reliability, sensitivity, and efficiency of the technique, while also reducing waste and environmental impact.

Emerging technologies and their potential impact

Emerging technologies have the potential to significantly impact the field of penetrant testing by improving the reliability, sensitivity, and efficiency of the technique. Some of the emerging technologies that could have a significant impact on the field include:

- Digital imaging: Advances in digital imaging technologies such as computed tomography (CT) and digital radiography (DR) are providing new ways to detect and analyze defects. These technologies can provide highly detailed images that can reveal defects that may not be visible with traditional penetrant testing methods.

- Robotics and automation: Robotic systems are being developed to automate the penetrant testing process. These systems can improve the accuracy and repeatability of the testing process and reduce the risk of operator error. They can also perform testing in areas that may be difficult or unsafe for human operators to access.
- Artificial intelligence (AI): AI and machine learning algorithms are being developed that can analyze images and data collected during penetrant testing. These algorithms can help detect defects more accurately and efficiently, reducing the risk of false positives and improving the overall reliability of the testing process.
- Smart materials: Smart materials are being developed that can change their properties in response to external stimuli. For example, researchers are developing materials that change color when exposed to a specific type of defect, making it easier to detect and analyze the defect.

Overall, these emerging technologies have the potential to improve the effectiveness and efficiency of penetrant testing, reduce the risk of errors, and improve the overall reliability of the technique. As these technologies continue to develop and become more widely available, they are likely to have a significant impact on the field of non-destructive testing.

Future trends and challenges in penetrant testing

The future trends in penetrant testing will likely involve advancements in technology, materials, and testing standards. Some of the future trends and challenges that could impact the field of penetrant testing include:

- Environmentally friendly materials: The development of more environmentally friendly penetrants, developers, and cleaners is an ongoing trend in the field. Many of the current materials used in penetrant testing are hazardous and require special handling and disposal. Developing more eco-friendly materials that are less harmful to the environment and human health is a priority for the industry.
- Automation and robotics: As mentioned earlier, the use of robotics and automation in penetrant testing is an emerging trend. These systems can improve accuracy, and repeatability, and reduce the risk of operator

error. In the future, we can expect to see more automation and robotics in the industry.

- Integration with other non-destructive testing methods: Penetrant testing is often used in conjunction with other non-destructive testing methods, such as radiography or ultrasonic testing. In the future, we can expect to see more integration of different non-destructive testing methods to improve the accuracy and reliability of defect detection.
- Advanced imaging and data analysis: Advances in digital imaging, artificial intelligence, and machine learning will likely improve the analysis of data collected during penetrant testing. This will allow for faster and more accurate defect detection, as well as a more detailed analysis of the data.
- Quality assurance and certification: As the industry continues to evolve, there will be an increased focus on quality assurance and certification. This will involve the development of new standards, guidelines, and training programs to ensure that operators are properly trained and qualified to perform penetrant testing.

Overall, the future of penetrant testing will likely involve the integration of new technologies and materials, as well as an increased focus on quality assurance and certification. The industry will continue to evolve and adapt to meet the changing needs of the marketplace and ensure that the technique remains a reliable and effective tool for defect detection.

CHAPTER IX

Chapter 9: Liquid Dye Penetrant Testing Q&A

Q: What is liquid dye penetrant testing?

A: Liquid dye penetrant testing is a non-destructive testing method used to detect surface defects in non-porous materials. It involves the application of a penetrant dye on the surface of the material, which is then allowed to seep into any surface cracks or defects. The excess dye is then removed, and a developer is applied, which draws the penetrant out of any defects, making them visible for inspection.

Q: What materials can be tested using liquid dye penetrant testing?

A: Liquid dye penetrant testing can be used on any non-porous material, including metals, plastics, ceramics, and composites.

Q: What are the steps involved in liquid dye penetrant testing?

A: The steps involved in liquid dye penetrant testing are as follows:

- Cleaning the surface to be tested
- Applying the penetrant dye
- Allowing the dye to seep into any surface defects
- Removing the excess dye
- Applying a developer to draw out the penetrant from any defects
- Inspecting the surface for defects

Q: What are the advantages of liquid dye penetrant testing?

A: The advantages of liquid dye penetrant testing include:

- It is a simple and inexpensive testing method
- It can be used on a wide range of materials
- It can detect small surface defects that may not be visible to the naked eye
- It does not require special training to perform

Q: What are the limitations of liquid dye penetrant testing?

A: The limitations of liquid dye penetrant testing include:

- It can only detect surface defects and not defects that are below the surface
- It is not suitable for use on porous materials
- It is not suitable for use on materials that are in service, as the penetrant dye can contaminate the material
- It may produce false positives if the surface being tested is not cleaned properly.

Q: What are the different types of penetrant dyes used in liquid dye penetrant testing?

A: There are three main types of penetrant dyes used in liquid dye penetrant testing:

- Fluorescent dye: This type of dye is visible under ultraviolet light and is typically used for high-sensitivity testing.
- Visible dye: This type of dye is visible under normal lighting conditions and is typically used for low-sensitivity testing.
- Dual-purpose dye: This type of dye can be used as either a fluorescent or visible dye, depending on the type of developer used.

Q: What are the safety precautions that should be taken when performing liquid dye penetrant testing?

A: The safety precautions that should be taken when performing liquid dye penetrant testing include:

- Wearing protective gloves and goggles to prevent skin and eye contact with the dye.
- Using the dye in a well-ventilated area to prevent inhalation of fumes.
- Storing the dye in a cool, dry place away from sources of heat and open flames.
- Disposing of used dye and developer in accordance with local regulations.

Q: What are the common applications of liquid dye penetrant testing?

A: Liquid dye penetrant testing is commonly used in a variety of industries, including aerospace, automotive, and manufacturing. It is used to detect surface defects in parts such as welds, castings, and forgings. It is also used to inspect components for cracks, porosity, and other surface defects before they are put into service.

Q: What are the factors that can affect the accuracy of liquid dye penetrant testing?

A: Several factors can affect the accuracy of liquid dye penetrant testing, including:

- Surface condition: The surface being tested should be clean and free of contaminants, as any residual oils, grease, or dirt can interfere with the penetrant dye's ability to seep into any surface defects.
- Penetrant dwell time: The amount of time the penetrant dye is left on the surface can affect the test's sensitivity. If the penetrant is left on the surface for too long, it can over-penetrate the defect and cause false positives.
- Developer application: The developer should be applied evenly to the surface and left on for the correct amount of time to allow any penetrant to be drawn out of any defects.
- Inspection environment: The inspection environment should be well-lit and free from any shadows or glare that can make it difficult to see any defects.
- Inspector experience: The inspector's experience and training can also affect the test's accuracy.

Q: What are the different levels of sensitivity in liquid dye penetrant testing?

A: There are three levels of sensitivity in liquid dye penetrant testing, as defined by industry standards:

- Level 1: This is the least sensitive level and is suitable for detecting larger defects.
- Level 2: This is a medium level of sensitivity and is suitable for detecting smaller defects.
- Level 3: This is the highest level of sensitivity and is suitable for detecting very small defects.

Q: How is the quality of liquid dye penetrant testing evaluated?

A: The quality of liquid dye penetrant testing can be evaluated by measuring the test's probability of detection (POD), which is a statistical measure of the test's ability to detect defects of a certain size. The POD can be calculated by performing a series of tests on known defects of varying sizes and comparing the results to the expected detection rate.

Q: What are the typical defects that can be detected using liquid dye penetrant testing?

A: Liquid dye penetrant testing can detect a variety of surface defects, including cracks, porosity, laps, seams, and lack of fusion. These defects can occur in a range of materials, including metals, plastics, ceramics, and composites.

Q: Can liquid dye penetrant testing be automated?

A: Yes, liquid dye penetrant testing can be automated using a variety of methods, including using robots or automated inspection equipment. Automated testing can help improve the consistency and accuracy of the test results and can be useful for inspecting large numbers of parts.

Q: What are the advantages of automated liquid dye penetrant testing?

A: The advantages of automated liquid dye penetrant testing include:

- Improved consistency and repeatability of the test results.
- Reduced operator error and fatigue.
- Increased testing speed, allowing for faster inspection of large numbers of parts.
- Improved safety, as automated systems, can reduce the need for manual handling of chemicals.

Q: What are the different types of developers used in liquid dye penetrant testing?

A: There are two main types of developers used in liquid dye penetrant testing: dry and wet. Dry developers are applied as a powder, while wet developers are applied as a liquid. Both types of developers are used to draw out the penetrant from any defects and create a visible indication for inspection. The choice of the developer will depend on the specific application and the requirements for sensitivity and ease of use.

Q: How does liquid dye penetrant testing compare to other non-destructive testing methods?

A: Liquid dye penetrant testing is one of several non-destructive testing methods used to detect surface defects. Other methods include magnetic particle testing, eddy current testing, and ultrasonic testing. Each method has its own advantages and limitations, and the choice of method will depend on the specific application and the requirements for sensitivity and ease of use. Liquid dye penetrant testing is generally a simple and inexpensive testing method, suitable for detecting small surface defects in non-porous materials.

Q: What are the limitations of liquid dye penetrant testing?

A: There are several limitations to liquid dye penetrant testing, including:

- It is only suitable for detecting surface defects and cannot detect internal defects or flaws.
- It is not suitable for use on porous materials or materials that absorb the penetrant dye.
- It requires a clean and dry surface for accurate results.
- It can be affected by surface finish, such as roughness or scratches, which can interfere with the penetrant dye's ability to seep into any defects.
- It requires a certain amount of operator skill and experience to ensure accurate results.

Q: What is the process for performing liquid dye penetrant testing?

A: The process for performing liquid dye penetrant testing typically includes the following steps:

- Clean the surface to be tested to remove any contaminants or debris.
- Apply the penetrant dye to the surface and allow it to dwell for a specified amount of time.
- Remove any excess penetrant dye from the surface.
- Apply the developer to the surface and allow it to dwell for a specified amount of time.
- Inspect the surface for any visible indications of defects.
- Evaluate the results and determine whether any defects are present.

Q: What are the different types of equipment used in liquid dye penetrant testing?

A: The equipment used in liquid dye penetrant testing typically includes:

- Penetrant spray system or applicator: used to apply the penetrant dye to the surface being tested.
- Developer applicator: used to apply the developer to the surface after the excess penetrant has been removed.

- UV light source: used to inspect the surface for fluorescent indications under ultraviolet light.
- Inspection booth: used to provide a controlled lighting environment for the inspection process.
- Cleaning and pre-treatment equipment: used to clean and prepare the surface before testing.
- Automated inspection equipment: used to automate the testing process and improve consistency and accuracy.

Q: What are the safety considerations for liquid dye penetrant testing?

A: The chemicals used in liquid dye penetrant testing can be hazardous if not handled properly. Safety considerations include:

- Wearing appropriate personal protective equipment, such as gloves, goggles, and aprons.
- Avoiding contact with the chemicals and ensuring adequate ventilation in the testing area.
- Properly disposing of used chemicals and following all applicable safety regulations and guidelines.
- Following proper handling and storage procedures for the chemicals and equipment.
- Conducting regular safety training for personnel involved in the testing process.

Q: How is the sensitivity of liquid dye penetrant testing determined?

A: The sensitivity of liquid dye penetrant testing is determined by the contrast between the penetrant dye and the developer. This contrast is affected by the type of penetrant dye used, the type of developer used, and the surface finish of the material being tested. Sensitivity can also be affected by the amount of time the penetrant dye is allowed to dwell on the surface and the amount of excess penetrant that is removed before the developer is applied. The sensitivity of the test can be evaluated by testing known defects of various sizes and shapes under controlled conditions.

Q: What are the advantages of using fluorescent penetrant dyes in liquid dye penetrant testing?

A: The advantages of using fluorescent penetrant dyes in liquid dye penetrant testing include:

- Improved sensitivity, as fluorescent dyes can be easier to detect under UV light than non-fluorescent dyes.
- Improved contrast, as fluorescent dyes, can provide a clearer indication of defects on the surface.
- Easier detection of small defects, as fluorescent dyes can be more effective at seeping into small cracks and flaws.
- Improved safety, as fluorescent dyes can be easier to see in low light conditions, reducing the need for bright lighting during the inspection.

Q: What are the advantages of using water-washable penetrant dyes in liquid dye penetrant testing?

A: The advantages of using water-washable penetrant dyes in liquid dye penetrant testing include:

- Reduced environmental impact, as water-washable penetrants can be easily washed off with water and do not require hazardous chemicals for removal.
- Improved safety, as water-washable penetrants are generally less hazardous than oil-based penetrants.
- Reduced cleaning time, as water-washable penetrants can be quickly and easily removed with water and do not require additional cleaning agents.
- Improved inspection accuracy, as water-washable penetrants can be less likely to leave residues that could interfere with the developer or inspection process.

Q: What is the difference between Type I and Type II penetrant testing methods?

A: Type I and Type II are two different categories of penetrant testing methods that are defined by the American Society for Testing and Materials (ASTM).

- Type I penetrant testing is a method that uses visible dye penetrants, which are visible to the naked eye under normal lighting conditions.
- Type I penetrants are typically used for detecting surface defects in materials that are not highly polished, such as rough castings or welds.
- Type II penetrant testing is a method that uses fluorescent dye penetrants, which are visible under ultraviolet (UV) light. Type II penetrants are typically used for detecting surface defects in materials

that have a highly polished surface finish, such as aircraft components or medical devices, where visible dye penetrants may not provide adequate contrast.

Q: What is the difference between the penetrant and developer used in liquid dye penetrant testing?

A:

- The penetrant and developer used in liquid dye penetrant testing are two different chemicals that work together to detect surface defects.
- The penetrant is a low-viscosity liquid that is applied to the surface being tested. The penetrant seeps into any surface defects, such as cracks or porosity, and is left on the surface for a specified amount of time to allow it to seep into any potential defects. The penetrant may be oil-based or water-based and may be visible or fluorescent, depending on the type of testing being performed.
- The developer is a high-viscosity, white powdery substance that is applied to the surface after the excess penetrant has been removed.
- The developer absorbs the penetrant from any defects, creating a visible indication of the defect. The developer may be applied by spraying or dusting and may contain additional ingredients to improve adhesion or sensitivity. The developer creates a visible contrast that makes it easier to identify defects on the surface.

Q: What are the limitations of liquid dye penetrant testing?

A: Liquid dye penetrant testing has several limitations that should be taken into consideration, including:

- Surface accessibility: The method relies on the penetrant being able to seep into surface defects, so surface accessibility is an important factor. Defects that are too deep or too narrow to be accessed by the penetrant may not be detectable.
- Limited depth of detection: Liquid dye penetrant testing is a surface method, so it is limited to detecting surface defects. It may not be effective at detecting defects that are below the surface.
- False positives: Some materials, such as porous materials or materials with a rough surface, may retain the penetrant even after cleaning, resulting in false positives.

- Potential for human error: The method requires a skilled inspector to properly apply and interpret the results. Errors in the application or interpretation of the results may result in false positives or false negatives.
- Limited detection of certain types of defects: The method is most effective at detecting surface defects, such as cracks or porosity, but may not be effective at detecting other types of defects, such as internal discontinuities or inclusions.

Q: What are the steps involved in performing liquid dye penetrant testing?

A: The steps involved in performing liquid dye penetrant testing typically include the following:

- Pre-cleaning: The surface being tested must be cleaned thoroughly to remove any surface contaminants that could interfere with the testing process. This may involve using a solvent cleaner or other cleaning agents.
- Application of penetrant: The penetrant is applied to the surface being tested and is allowed to seep into any surface defects for a specified amount of time, which can range from several minutes to several hours.
- Excess penetrant removal: After the specified dwell time has passed, the excess penetrant is removed from the surface using a solvent cleaner or emulsifier. This step is important to ensure that only the penetrant that has seeped into any defects remains on the surface.
- Application of developer: The developer is applied to the surface and is allowed to dry for a specified amount of time, typically ranging from several minutes to several hours.
- Inspection: The surface is inspected visually, either under normal lighting or under UV light, depending on the type of penetrant used. The inspector looks for any indications of defects, which will be visible as a bright red or fluorescent indication against the white developer.
- Post-cleaning: After the inspection is complete, the surface is cleaned to remove any remaining penetrant or developer. The cleaning process may involve using a solvent cleaner or other cleaning agents.

Q: What are some common applications of liquid dye penetrant testing?

A: Liquid dye penetrant testing is commonly used in a variety of industries and applications where surface defects may be a concern. Some common applications of liquid dye penetrant testing include:

- Aerospace: Liquid dye penetrant testing is used to inspect critical aerospace components, such as engine components, landing gear, and structural components.
- Automotive: Liquid dye penetrant testing is used to inspect automotive parts, such as engine blocks, transmission housings, and brake components.
- Manufacturing: Liquid dye penetrant testing is used to inspect a variety of manufactured components, including welds, castings, and forgings.
- Oil and gas: Liquid dye penetrant testing is used to inspect pipelines, storage tanks, and other components in the oil and gas industry.
- Power generation: Liquid dye penetrant testing is used to inspect components in power generation facilities, including turbines, generators, and boilers.
- Railways: Liquid dye penetrant testing is used to inspect railway components, such as axles, wheels, and other critical components.
- Welding: Liquid dye penetrant testing is used to inspect welded components for surface cracks or other defects.

Q: What are the advantages of using liquid dye penetrant testing?

A: There are several advantages of using liquid dye penetrant testing, including:

- Versatility: Liquid dye penetrant testing can be used on a variety of materials, including metals, plastics, and ceramics.
- Cost-effective: Liquid dye penetrant testing is a relatively low-cost method compared to other non-destructive testing methods.
- Non-destructive: Liquid dye penetrant testing is a non-destructive testing method, which means that it does not damage the component being tested.
- High sensitivity: Liquid dye penetrant testing can detect very small surface defects, making it an effective method for inspecting critical components.
- Easy to perform: Liquid dye penetrant testing can be performed relatively easily, with minimal training required for the operator.

- Portable: Liquid dye penetrant testing can be performed on-site, making it a useful method for field inspections.
- Regulatory compliance: Liquid dye penetrant testing is a widely accepted and recognized testing method, and is often required by regulatory agencies in certain industries.

Q: What are some common types of penetrants used in liquid dye penetrant testing?

A: There are several types of penetrants used in liquid dye penetrant testing, including:

- Water-washable penetrants: These penetrants can be easily removed from the surface using water or a water-based cleaner. They are commonly used in applications where water-based cleaning is acceptable, such as in the automotive industry.
- Solvent-removable penetrants: These penetrants are removed using a solvent cleaner or emulsifier. They are commonly used in applications where water-based cleaning is not acceptable, such as in the aerospace industry.
- Post-emulsifiable penetrants: These penetrants are designed to be removed using a post-emulsifier after the excess penetrant has been removed from the surface. They are commonly used in applications where higher sensitivity is required, such as in the nuclear industry.
- Lipophilic penetrants: These penetrants are designed for use on non-porous materials, such as plastics and ceramics. They contain a dye that is soluble in oil, allowing it to seep into surface defects in non-porous materials.
- Hydrophilic penetrants: These penetrants are designed for use on porous materials, such as castings and welds. They contain a dye that is soluble in water, allowing it to seep into surface defects in porous materials.

Q: What are the different steps involved in liquid dye penetrant testing?

A: The liquid dye penetrant testing process typically involves the following steps:

- Pre-cleaning: The component being tested is cleaned to remove any surface contaminants or coatings that may interfere with the penetrant.

This step is critical for the accuracy of the test.

- Application of penetrant: The penetrant is applied to the surface of the component, either by spraying or brushing. The penetrant is allowed to soak into any surface defects for a specified amount of time, which can range from a few minutes to several hours.
- Excess penetrant removal: The excess penetrant is removed from the surface of the component using a cleaner or solvent. This step is important to ensure that only the penetrant that has seeped into surface defects is visible in the subsequent steps.
- Application of developer: The developer is applied to the surface of the component. The developer draws the penetrant out of any surface defects and spreads it out over a larger area, making it more visible.
- Inspection: The component is inspected under appropriate lighting conditions to identify any visible penetrant indications, which may indicate the presence of a surface defect.
- Post-cleaning: The component is cleaned to remove any remaining penetrant or developer. This step is important to ensure that the surface of the component is left clean and free of any residues that may interfere with subsequent operations or inspections.

Q: What are some factors that can affect the accuracy of liquid dye penetrant testing?

A: There are several factors that can affect the accuracy of liquid dye penetrant testing, including:

- Surface preparation: The surface of the component being tested must be properly prepared to ensure that it is free from any contaminants or coatings that may interfere with the penetrant. Any surface preparation that is not carried out correctly can lead to false indications.
- Penetrant selection: The selection of the appropriate type of penetrant for the application is critical to ensure that the test is accurate. The penetrant selected should be compatible with the material being tested, and should be able to penetrate any surface defects present.
- Penetrant application: The penetrant should be applied to the surface of the component according to the manufacturer's recommendations. Overapplication or underapplication of the penetrant can lead to inaccurate results.

- Soak time: The amount of time that the penetrant is left to soak into surface defects is critical to the accuracy of the test. The soak time should be long enough to allow the penetrant to seep into any surface defects, but not so long that the penetrant begins to dry on the surface.
- Developer application: The developer should be applied to the surface of the component according to the manufacturer's recommendations. Overapplication or underapplication of the developer can lead to inaccurate results.
- Lighting conditions: Proper lighting conditions are critical for the accurate detection of penetrant indications. The inspection area should be well-lit, and the lighting should be appropriate for the type of penetrant being used.
- Operator experience: The experience of the operator can also affect the accuracy of the test. Proper training and experience are important for the operator to be able to accurately interpret the results of the test.

Q: What are some advantages of liquid dye penetrant testing?
A: Some advantages of liquid dye penetrant testing include:

- Versatility: Liquid dye penetrant testing can be used on a wide variety of materials, including metals, plastics, ceramics, and composites. It can be used to detect surface defects such as cracks, porosity, laps, and seams.
- Sensitivity: Liquid dye penetrant testing can detect very small surface defects that may not be visible to the naked eye. This makes it a highly sensitive inspection method that can be used to detect even very small defects.
- Cost-effectiveness: Liquid dye penetrant testing is a relatively inexpensive inspection method when compared to other methods such as X-ray or ultrasonic testing. This makes it a popular choice for many applications where cost is a concern.
- Ease of use: Liquid dye penetrant testing is a relatively simple inspection method that can be easily carried out by trained personnel. This makes it a popular choice for many industries where the inspection of components is a routine operation.
- Portable: Liquid dye penetrant testing can be carried out in the field, making it a popular choice for many applications where components cannot be easily transported to an inspection facility.

- Non-destructive: Liquid dye penetrant testing is a non-destructive inspection method, which means that the component being tested is not damaged in any way during the inspection process. This makes it a popular choice for many industries where the integrity of the component must be maintained.

Q: What are some limitations of liquid dye penetrant testing?
A: Some limitations of liquid dye penetrant testing include:

- Surface accessibility: Liquid dye penetrant testing is only effective for surface defects that are accessible to the penetrant. It cannot detect defects that are located beneath the surface of the component.
- Material compatibility: The penetrant used for liquid dye penetrant testing may not be compatible with all materials. In some cases, the penetrant may react with the material being tested, which can result in false indications or damage to the component.
- Surface finish: The surface finish of the component being tested can affect the accuracy of the test. Components with a rough surface finish may be more difficult to clean, which can lead to false indications. In addition, the developer may settle into surface imperfections, making it more difficult to identify penetrant indications.
- Limited detection of certain types of defects: Liquid dye penetrant testing is primarily used to detect surface defects such as cracks, porosity, laps, and seams. It is not effective for detecting other types of defects such as corrosion, erosion, or fatigue cracking.
- Operator interpretation: The accuracy of the test can be affected by the experience and training of the operator. The interpretation of the results is subjective, and a less experienced operator may miss or misinterpret indications, leading to false results.

Q: What are some other limitations of liquid dye penetrant testing?
A: Other limitations of liquid dye penetrant testing include:

- Time-consuming: Liquid dye penetrant testing requires several steps, including surface preparation, application of the penetrant, and development of the indications. This can make the testing process time-consuming and may not be practical for high-volume production lines.

- Sensitivity limitations: While liquid dye penetrant testing can detect very small surface defects, it may not be sensitive enough to detect defects that are very shallow or very narrow. In addition, some types of defects, such as fatigue cracks, may not produce indications that are easily detectable with liquid penetrant testing.
- Health and safety considerations: The chemicals used in liquid dye penetrant testing can be hazardous to the health of the operator if not used properly. These chemicals may require special handling and disposal procedures, and appropriate personal protective equipment must be worn.
- Environmental considerations: The chemicals used in liquid dye penetrant testing can be harmful to the environment if not handled and disposed of properly. Appropriate measures must be taken to prevent these chemicals from entering the environment.
- Limited ability to distinguish between different types of defects: Liquid dye penetrant testing can only detect the presence of a defect, but it cannot distinguish between different types of defects. For example, it cannot differentiate between a crack and a porosity indication.
- Inability to provide information about the depth of the defect: Liquid dye penetrant testing can only detect surface defects and cannot provide information about the depth of the defect. This can be a limitation when determining the severity of the defect and the appropriate course of action.

Q: What are some safety considerations for liquid dye penetrant testing?
A: Some safety considerations for liquid dye penetrant testing include:

- Personal protective equipment: Appropriate personal protective equipment (PPE) should be worn when handling the chemicals used in liquid dye penetrant testing. This may include gloves, safety glasses, and protective clothing.
- Ventilation: Liquid dye penetrant testing should be carried out in a well-ventilated area to prevent the buildup of fumes and to ensure that the operator has a continuous supply of fresh air.
- Chemical handling: The chemicals used in liquid dye penetrant testing can be hazardous if not handled properly. Operators should receive appropriate training on how to handle and store these chemicals, and appropriate safety measures should be in place to prevent spills and

leaks.

- Disposal: The chemicals used in liquid dye penetrant testing must be disposed of properly to prevent harm to the environment. Operators should be trained on how to properly dispose of these chemicals and should follow local regulations and guidelines.
- Emergency procedures: Emergency procedures should be in place in case of spills or accidents. These procedures should include how to handle spills, how to clean up the chemicals, and how to respond to injuries or illnesses caused by exposure to the chemicals.
- Training: Operators should receive appropriate training on how to carry out liquid dye penetrant testing safely. This training should include information on how to handle the chemicals, how to use PPE, and how to respond to emergency situations.

Q: What are some advantages of liquid dye penetrant testing?
A: Some advantages of liquid dye penetrant testing include:

- Versatility: Liquid dye penetrant testing can be used to detect surface defects in a variety of materials, including metals, plastics, ceramics, and composites.
- Cost-effectiveness: Liquid dye penetrant testing is a relatively inexpensive testing method, making it accessible to a wide range of industries and applications.
- Sensitivity: Liquid dye penetrant testing can detect very small surface defects, including cracks and porosity, that may not be visible to the naked eye.
- Reliability: When performed correctly, liquid dye penetrant testing can be a very reliable method for detecting surface defects. The indications produced are visible and easily identifiable, and false indications are rare.
- Ease of use: Liquid dye penetrant testing is a relatively simple and straightforward testing method, requiring only basic training and equipment.
- Non-destructive: Liquid dye penetrant testing is a non-destructive testing method, meaning that it does not damage the component being tested. This makes it a valuable tool for inspecting components that cannot be replaced or repaired easily.
- Wide acceptance: Liquid dye penetrant testing is widely accepted in many industries, including aerospace, automotive, and manufacturing.

This means that there are established standards and procedures for conducting the test, and the results are widely recognized and accepted.

Q: What are some limitations of liquid dye penetrant testing?
A: Some limitations of liquid dye penetrant testing include:

- Surface defects only: Liquid dye penetrant testing is only effective at detecting surface defects, and cannot be used to detect defects that are located beneath the surface of a material.
- Time-consuming: The process of liquid dye penetrant testing can be time-consuming, as it involves several steps and requires sufficient time for the penetrant to soak into any surface defects.
- Sensitivity to surface condition: The effectiveness of liquid dye penetrant testing can be affected by the surface condition of the material being tested. If the surface is rough or uneven, it may be more difficult for the penetrant to flow into any surface defects.
- Discoloration of the material: The penetrant used in liquid dye penetrant testing can stain or discolor the material being tested, which may be an issue in some applications where the appearance of the material is important.
- Operator-dependent: The reliability of liquid dye penetrant testing can be affected by the skill and experience of the operator. If the operator is not properly trained or does not follow the correct procedures, the results of the test may be inaccurate.
- Limited to certain materials: Liquid dye penetrant testing may not be effective for all materials, as some materials may not be able to hold the penetrant or may be too porous to provide reliable results.
- Limited to accessible areas: Liquid dye penetrant testing can only be used in areas that are accessible for the application and removal of the penetrant. In some cases, this may limit its effectiveness or applicability.

Q: What are the steps involved in liquid dye penetrant testing?
A: The steps involved in liquid dye penetrant testing typically include:

- Pre-cleaning: The surface of the material being tested is first cleaned to remove any dirt, oil, or other contaminants that may interfere with the test. This can be done using solvents, detergents, or other cleaning agents.

- Penetrant application: A liquid penetrant is then applied to the surface of the material, either by spraying, brushing, or immersion. The penetrant is left on the surface for a specified period of time to allow it to seep into any surface defects.
- Dwell time: The dwell time is the period during which the penetrant is left on the surface of the material. The length of the dwell time depends on the type of penetrant being used and the size and depth of the defects being sought.
- Removal of excess penetrant: After the dwell time has elapsed, the excess penetrant is removed from the surface of the material using a solvent, emulsifier, or water rinse.
- Application of developer: A developer is then applied to the surface of the material to draw out any penetrant that has seeped into surface defects. The developer may be a dry powder, a wet suspension, or a water-soluble liquid.
- Dwell time: The developer is left on the surface of the material for a specified period of time to allow it to draw out any penetrant that has seeped into surface defects.
- Inspection: The surface of the material is then inspected for indications of defects. Indications will appear as a bright, contrasting color against the background color of the developer.
- Post-cleaning: Finally, the surface of the material is cleaned to remove any residual penetrant or developer, and to restore the material to its original condition.

Q: What are some advantages of liquid dye penetrant testing?
A: Some advantages of liquid dye penetrant testing include:

- Non-destructive: Liquid dye penetrant testing is a non-destructive method of testing, meaning that it does not damage the material being tested.
- Low cost: Compared to other non-destructive testing methods, such as x-ray or ultrasound testing, liquid dye penetrant testing is relatively inexpensive.
- Versatility: Liquid dye penetrant testing can be used to test a variety of different materials, including metals, plastics, ceramics, and composites.
- High sensitivity: Liquid dye penetrant testing is highly sensitive to surface defects, even those that are not visible to the naked eye.

- Simple to use: Liquid dye penetrant testing is a relatively simple and straightforward process, and does not require specialized equipment or extensive training.
- Fast results: Liquid dye penetrant testing can provide rapid results, with indications of defects appearing within a few minutes of the application of the developer.
- Portable: Liquid dye penetrant testing can be performed on-site, making it a convenient and practical method of testing for many applications.

Q: What are the limitations of liquid dye penetrant testing?
A: Some limitations of liquid dye penetrant testing include:

- Surface defects only: Liquid dye penetrant testing is only effective at detecting surface defects, and may not detect defects that are below the surface of the material.
- Limited to non-porous materials: Liquid dye penetrant testing is only effective on non-porous materials, as the penetrant cannot seep into porous materials.
- Limited to specific defect types: Liquid dye penetrant testing is most effective at detecting surface-breaking defects, such as cracks, laps, and porosity. It may not be effective at detecting other types of defects, such as internal voids or inclusions.
- Limited to certain materials: While liquid dye penetrant testing can be used on a wide range of materials, it may not be suitable for certain types of materials, such as those that are highly porous or have a rough or irregular surface.
- Limited temperature range: Liquid dye penetrant testing is typically only effective within a certain temperature range, and may not work effectively at very high or very low temperatures.
- Environmental concerns: The penetrants and developers used in liquid dye penetrant testing can be harmful to the environment, and may require special handling and disposal procedures.
- Human error: Liquid dye penetrant testing is a manual process and is subject to human error, such as incomplete cleaning or inadequate application of the penetrant or developer.

CHAPTER X

Chapter 10: Penetrant Testing Important Interview Q&A

What is penetrant testing and how does it differ from other non-destructive testing methods?

Answer: Penetrant testing is a non-destructive testing method that involves applying a liquid penetrant to the surface of a material, which is then drawn into any surface-breaking defects by capillary action. The excess penetrant is then removed and a developer is applied, which helps to pull the penetrant out of the defects, making them visible to the inspector. This method is different from other NDT methods such as radiographic testing, ultrasonic testing, and magnetic particle testing, which use different physical principles to detect defects.

What types of materials can be tested using penetrant testing?

Answer: Penetrant testing can be used on a wide variety of materials, including metals, plastics, ceramics, and composites. However, it is not suitable for use on porous materials or on materials that cannot withstand the cleaning and drying processes required before and after the penetrant is applied.

What are the different types of penetrants that can be used in penetrant testing?

Answer: There are two main types of penetrants: visible and fluorescent. Visible penetrants are colored and can be seen with the naked eye, while fluorescent penetrants emit light when exposed to ultraviolet light and can only be seen with the aid of a UV lamp.

What are the advantages and disadvantages of penetrant testing?

Answer: Advantages of penetrant testing include its cost-effectiveness, ease of use, and ability to detect surface-breaking defects. It also does not require any special equipment. The disadvantages include the need for cleaning and drying of the surface, the requirement of a dark room for fluorescent penetrant testing, and the possibility of false indications if the surface is not properly cleaned.

What are the steps involved in the penetrant testing process?

Answer: The general steps involved in the penetrant testing process are:

- -Cleaning the surface of the part to be tested to remove any contaminants
- -Applying the penetrant to the surface
- -Allowing the penetrant to dwell on the surface for a specified amount of time
- -Removing the excess penetrant from the surface
- -Applying a developer to the surface
- -Inspecting the surface for indications of defects.

How do you interpret the results of a penetrant testing inspection?

Answer: The results of a penetrant testing inspection are interpreted by visually inspecting the surface of the material for indications of defects. These indications can appear as stains, color variations, or other visible changes on the surface. The inspector will then evaluate the size, shape, location, and number of indications to determine if they are likely to be defects and if they require further evaluation or repair.

How do you ensure the quality of penetrant testing results?

Answer: To ensure the quality of penetrant testing results, it is important to follow proper inspection procedures and to use qualified personnel to perform the inspection. It is also important to use properly calibrated equipment and to maintain strict control over the penetrant and developer materials used in the inspection. Additionally, it is important to follow up on any indications with further testing or evaluation to confirm their nature.

How do you ensure the safety of personnel during penetrant testing?

Answer: To ensure the safety of personnel during penetrant testing, it is important to use protective equipment such as gloves and safety glasses to prevent contact with penetrant materials. It is also important to properly ventilate the area in which the inspection is taking place and to follow proper disposal procedures for any used penetrant or developer materials.

What are some common penetrant testing applications?

Answer: Common penetrant testing applications include:

- -Inspection of aircraft components such as gears and landing gear parts
- -Inspection of pressure vessels and pipelines
- -Inspection of castings and forgings
- -Inspection of welds
- -Inspection of gears, bearings, and other mechanical components

How do you stay current with the latest penetrant testing techniques and technologies?

Answer: To stay current with the latest penetrant testing techniques and technologies, one can attend training and certification courses offered by organizations such as the American Society for Nondestructive Testing (ASNT), read industry publications, and participate in professional organizations such as the International Association of Penetrant Testing Professionals (IAPTP). It's also beneficial to stay informed about changes in industry standards and regulations.

How do you ensure that the penetrant testing equipment is calibrated and in good working condition?

Answer: To ensure that penetrant testing equipment is calibrated and in good working condition, it is important to follow the manufacturer's recommended calibration procedures and to maintain a schedule for regular calibration and maintenance checks. This may include checking the calibration of UV lamps, measuring the intensity of the light emitted, and inspecting the equipment for any signs of wear or damage. It is also important to keep accurate records of the equipment's calibration and maintenance status.

How do you select the appropriate penetrant and developer for a particular application?

Answer: To select the appropriate penetrant and developer for a particular application, it is important to consider the type of material being inspected, the size and location of the defects, and the lighting conditions of the inspection area. For example, a fluorescent penetrant would be more appropriate for a dark inspection area, while a visible penetrant would be more appropriate for a well-lit inspection area. It is also important to consider the compatibility of the penetrant and developer with the material being inspected.

How do you properly store and handle penetrant and developer materials?

Answer: To properly store and handle penetrant and developer materials, it is important to follow the manufacturer's instructions for storage and handling. This may include keeping the materials in a cool, dry place, away from heat and light sources, and in properly labeled containers. It's also important to follow safety precautions such as wearing gloves and safety glasses when handling the materials.

How do you evaluate and document the results of a penetrant testing inspection?

Answer: To evaluate and document the results of a penetrant testing inspection, the inspector should visually inspect the surface of the material for indications of defects and evaluate the size, shape, location, and number of indications. The inspector should also take photographs or other forms of documentation of the indications and provide a written report describing the inspection results and any recommendations for further evaluation or repair.

How do you handle false indications during a penetrant testing inspection?

Answer: False indications can occur during a penetrant testing inspection due to surface contaminants, improper cleaning or preparation of the surface, or other factors. To handle false indications, the inspector should re-clean the surface and reapply the penetrant and developer to confirm the presence of an actual defect. If the indication persists, the inspector should evaluate the indication to determine if it is likely to be a defect or a false indication and document the results accordingly. If the indication is determined to be a false indication, the inspector should take steps to eliminate the cause of the false indication in order to prevent it from occurring in the future.

How do you handle parts that have been painted or coated?

Answer: When inspecting parts that have been painted or coated, it is important to remove the coating in order to expose the surface of the material to the penetrant. This can be done using chemical or mechanical methods, such as using a solvent or abrasive blasting. Once the coating has been removed, the surface should be cleaned and inspected in accordance with the standard penetrant testing procedures.

What are the main industry standards and regulations related to penetrant testing?

Answer: The main industry standards and regulations related to penetrant testing include:

- -ASTM E1417 - Standard Practice for Liquid Penetrant Testing
- -ASME Section V - Nondestructive Examination
- -NAS410 - Nondestructive Inspection (NDI) for Aerospace
- -ISO 3452-2 - Non-destructive testing - Penetrant testing and magnetic particle testing - Part 2: Penetrant testing

- -ASNT-TC-1A - Recommended Practice for Personnel Qualification and Certification in Nondestructive Testing.

It's important to be familiar with these standards and regulations and to follow them in order to ensure the quality and accuracy of penetrant testing results.

How do you ensure that the penetrant testing inspection is performed in accordance with the applicable industry standards and regulations?

Answer: To ensure that the penetrant testing inspection is performed in accordance with the applicable industry standards and regulations, it is important to follow the established inspection procedures and guidelines provided in the standards. This includes following the proper cleaning, penetrant application, and developer application procedures, as well as adhering to the recommended inspection methods and acceptance criteria. It is also important to use qualified personnel who are trained and certified in accordance with the standards and to maintain accurate records of the inspection results and any deviations from the established procedures.

How do you train and certify personnel in penetrant testing?

Answer: To train and certify personnel in penetrant testing, it is important to provide comprehensive training in the theory and practical application of penetrant testing, including instruction on the proper use of equipment, inspection procedures, and interpretation of results. It is also important to provide hands-on training and practical experience under the supervision of a qualified instructor. After completing the training, personnel should be evaluated through a certification process which typically includes a written exam, practical exam, and review of their work experience. Organizations such as the American Society for Nondestructive Testing (ASNT) and the International Association of Penetrant Testing Professionals (IAPTP) provide certifications for penetrant testing.

What is the importance of proper record-keeping for penetrant testing?

Answer: Proper record-keeping for penetrant testing is important to ensure that the inspection results are accurate and can be easily referenced in the future. It also helps to demonstrate compliance with industry standards and regulations. The records should include information such as the inspection results, the personnel who performed the inspection, the equipment used, and any deviations from the established procedures. These records should be kept in a secure location, and they should be easily accessible to authorized personnel.

How do you handle penetrant testing on large or complex parts?

Answer: Handling penetrant testing on large or complex parts can be challenging due to the difficulty of accessing all areas of the part and the need for proper handling and support of the part during the inspection. To handle this, it is important to have a well-defined inspection plan that includes the proper positioning and handling of the part, as well as the use of appropriate equipment and personnel. It may also be necessary to divide the inspection into smaller, manageable areas, and to use specialized inspection techniques such as remote-controlled cameras and robotic inspection systems to access hard-to-reach areas.

How do you handle penetrant testing of parts with tight tolerances?

Answer: Handling penetrant testing of parts with tight tolerances can be challenging due to the need for precise inspection and the potential for damage to the part during the inspection process. To handle this, it is important to use a low-viscosity penetrant that can easily flow into tight tolerance areas without leaving residue. It is also important to use a developer that will not damage the part or leave residue. Additionally, the inspection should be done using a magnifying lens or microscope to ensure precise inspection, and the inspector should be highly skilled and experienced in the inspection of parts with tight tolerances.

How do you handle penetrant testing of parts with complex geometries?

Answer: Handling penetrant testing of parts with complex geometries can be challenging due to the difficulty of accessing all areas of the part and the need for proper handling and support of the part during the inspection. To handle this, it is important to have a well-defined inspection plan that includes the proper positioning and handling of the part, as well as the use of appropriate equipment and personnel. It may also be necessary to use specialized inspection techniques such as computed tomography (CT) scanning or 3D scanning to create a digital model of the part and to identify potential defects.

How do you handle penetrant testing of parts with a large surface area?

Answer: Handling penetrant testing of parts with a large surface area can be challenging due to the time and resources required to cover the entire surface. To handle this, it is important to have a well-defined inspection plan that includes the proper positioning and handling of the part, as well as the use of appropriate equipment and personnel. It may also be necessary to divide the inspection into smaller, manageable areas, and to use specialized inspection techniques such as remote-controlled cameras or drones to cover a larger area more efficiently. Additionally, it's important to properly plan and schedule the inspection in order to minimize disruptions to production or other operations.

What are the key factors to consider when selecting a penetrant testing service provider?

Answer: When selecting a penetrant testing service provider, it is important to consider the following key factors:

- The provider's experience and expertise in penetrant testing
- -The provider's certifications and compliance with industry standards and regulations
- -The provider's equipment and facilities
- -The provider's flexibility and ability to adapt to the specific needs of your project
- -The provider's reputation and references
- -The provider's pricing and cost-effectiveness
- -The provider's ability to provide accurate and timely results and reporting

By considering these key factors, you can be confident that you are selecting a qualified and reputable service provider that can meet your specific needs and deliver accurate, reliable results.

How do you ensure the security and confidentiality of the penetrant testing results?

Answer: To ensure the security and confidentiality of the penetrant testing results, it is important to have strict controls in place to limit access to the inspection results. This may include using password-protected software to store and share the results, or storing the results in a secure physical location. It's also important to have a clear policy in place to control access to the results and to communicate this policy to all personnel involved in the inspection process. Additionally, it is important to have a clear and specific confidentiality agreement with the service provider, which outlines the terms of confidentiality and responsibility for maintaining the security and confidentiality of results.

How do you evaluate the effectiveness of a penetrant testing program?

Answer: To evaluate the effectiveness of a penetrant testing program, it is important to track and analyze data related to the inspection process, including the number of defects detected, the number of false indications, the length of time required for the inspection, and the cost of the inspection. It's also important to track and analyze data related to the personnel and equipment involved in the inspection process. Additionally, by evaluating customer feedback and measuring the performance of the parts that have been tested can help to provide a more complete picture of the program's effectiveness. By evaluating this data, it is possible to identify areas for improvement and to make adjustments to the program as needed in order to optimize its effectiveness.

How do you handle penetrant testing of parts with high-temperature applications?

Answer: Handling penetrant testing of parts with high-temperature applications can be challenging due to the need to maintain the integrity of the part while performing the inspection. To handle this, it is important to use a penetrant and developer that can withstand the high temperatures without breaking down or losing effectiveness. Additionally, it is important to use proper cooling methods or protective equipment to prevent overheating of the part during the inspection process. It is also important to use a qualified inspector who has experience in inspecting high-temperature parts and to follow proper safety procedures to ensure the safety of the personnel involved.

How do you handle penetrant testing of parts with low-temperature applications?

Answer: Handling penetrant testing of parts with low-temperature applications can be challenging due to the need to maintain the integrity of the part while performing the inspection. To handle this, it is important to use a penetrant and developer that can withstand the low temperatures without freezing or losing effectiveness. Additionally, it is important to use proper heating methods or protective equipment to prevent freezing of the part during the inspection process.

It is also important to use a qualified inspector who has experience in inspecting low-temperature parts and to follow proper safety procedures to ensure the safety of the personnel involved.

How do you handle penetrant testing of parts with hazardous materials?

Answer: Handling penetrant testing of parts with hazardous materials can be challenging due to the need to maintain the integrity of the part while performing the inspection, as well as the need to protect the safety of the personnel involved. To handle this, it is important to use a penetrant and developer that are compatible with the hazardous materials and that do not pose a risk to the safety of the personnel.

Additionally, it is important to use proper protective equipment and to follow proper safety procedures to ensure the safety of the personnel involved. It is also important to use a qualified inspector who has experience in inspecting parts with hazardous materials and to be aware of the disposal procedures of the materials used in the inspection process.

How do you handle penetrant testing of parts with intricate internal geometries?

Answer: Handling penetrant testing of parts with intricate internal geometries can be challenging due to the difficulty of accessing all areas of the part and the need for proper handling and support of the part during the inspection. To handle this, it is important to have a well-defined inspection plan that includes the proper positioning and handling of the part, as well as the use of appropriate equipment and personnel.

It may also be necessary to use specialized inspection techniques such as borescopy, endoscopy, or radiography to access the internal geometries of the part. Additionally, it is important to use a penetrant with a low viscosity that can easily flow into tight spaces without leaving residue.

How do you handle penetrant testing of parts with irregular shapes?

Answer: Handling penetrant testing of parts with irregular shapes can be challenging due to the difficulty of maintaining proper contact between the penetrant and the surface of the part during the inspection. To handle this, it is important to have a well-defined inspection plan that includes the proper positioning and handling of the part, as well as the use of appropriate equipment and personnel.

It may also be necessary to use specialized inspection techniques such as robotic inspection systems or 3D scanning to ensure that all areas of the part are properly inspected. Additionally, it is important to use a penetrant with a low viscosity that can easily conform to the irregular shape of the part without leaving residue.

How do you handle penetrant testing of parts in a production environment?

Answer: Handling penetrant testing of parts in a production environment can be challenging due to the need to minimize disruptions to production and the need to ensure accurate and timely inspection results. To handle this, it is important to have a well-defined inspection plan that includes the proper positioning and handling of the part, as well as the use of appropriate equipment and personnel.

It may also be necessary to use specialized inspection techniques such as inline inspection systems that can inspect parts while they are still in the production process. Additionally, it is important to properly plan and schedule the inspection in order to minimize disruptions to production and to have a system in place for quickly and efficiently evaluating and documenting the inspection results.

How do you handle penetrant testing of parts that have been exposed to extreme environmental conditions?

Answer: Handling penetrant testing of parts that have been exposed to extreme environmental conditions can be challenging due to the potential damage to the surface of the part, and the need to ensure accurate and reliable inspection results. To handle this, it is important to have a well-defined inspection plan that includes the proper positioning and handling of the part, as well as the use of appropriate equipment and personnel.

Before the inspection, it's important to evaluate the condition of the part, and to determine whether any cleaning, surface preparation, or conditioning is necessary to ensure accurate results. It may also be necessary to use specialized inspection techniques such as thermal imaging or infrared cameras to evaluate the condition of the part. Additionally, it is important to have a system in place for quickly and efficiently evaluating and documenting the inspection results.

How do you handle penetrant testing of parts with a low surface finish?

Answer: Handling penetrant testing of parts with a low surface finish can be challenging due to the difficulty of maintaining proper contact between the penetrant and the surface of the part during the inspection. To handle this, it is important to have a well-defined inspection plan that includes the proper positioning and handling of the part, as well as the use of appropriate equipment and personnel. It may also be necessary to use specialized surface preparation techniques such as abrasive blasting or chemical etching to improve the surface finish of the part before the inspection. Additionally, it is important to use a penetrant with a low viscosity that can easily conform to the surface of the part without leaving residue.

How do you handle penetrant testing of parts with a high surface finish?

Answer: Handling penetrant testing of parts with a high surface finish can be challenging due to the difficulty of maintaining proper contact between the penetrant and the surface of the part during the inspection. To handle this, it is important to have a well-defined inspection plan that includes the proper positioning and handling of the part, as well as the use of appropriate equipment and personnel.

It may also be necessary to use specialized surface preparation techniques such as polishing or lapping to improve the surface finish of the part before the inspection. Additionally, it is important to use a penetrant with a low viscosity that can easily conform to the surface of the part without leaving residue.

How do you handle penetrant testing of parts with complex internal geometry?

Answer: Handling penetrant testing of parts with complex internal geometry can be challenging due to the difficulty of accessing all areas of the part and the need for proper handling and support of the part during the inspection. To handle this, it is important to have a well-defined inspection plan that includes the proper positioning and handling of the part, as well as the use of appropriate equipment and personnel.

It may also be necessary to use specialized inspection techniques such as borescopy, endoscopy, or radiography to access the internal geometries of the part. Additionally, it is important to use a penetrant with a low viscosity that can easily flow into tight spaces without leaving residue.

How do you handle penetrant testing of parts with tight tolerance?

Answer: Handling penetrant testing of parts with tight tolerance can be challenging due to the need for precise inspection and the potential for damage to the part during the inspection process. To handle this, it is important to use a low viscosity penetrant that can easily flow into tight tolerance areas without leaving residue. It is also important to use a developer that will not damage the part or leave residue.

Additionally, the inspection should be done using a magnifying lens or microscope to ensure precise inspection, and the inspector should be highly skilled and experienced in the inspection of parts with tight tolerances.

How do you handle penetrant testing of parts with tight tolerance and complex geometries?

Answer: Handling penetrant testing of parts with tight tolerance and complex geometries can be a complex and challenging task. To handle this, it is important to have a well-defined inspection plan that includes the proper positioning and handling of the part, as well as the use of appropriate equipment and personnel.

It may also be necessary to use specialized inspection techniques such as computed tomography (CT) scanning or 3D scanning to create a digital model of the part and to identify potential defects.

Additionally, it is important to use a low viscosity penetrant that can easily flow into tight tolerance areas without leaving residue and a developer that will not damage the part or leave residue. The inspection should be done by a highly skilled and experienced inspector using magnifying lens or microscope to ensure precise inspection.

How do you handle penetrant testing of parts that have been coated or painted?

Answer: Handling penetrant testing of parts that have been coated or painted can be challenging due to the difficulty of maintaining proper contact between the penetrant and the surface of the part during the inspection. To handle this, it is important to have a well-defined inspection plan that includes the proper positioning and handling of the part, as well as the use of appropriate equipment and personnel. Before the inspection, it's important to evaluate the condition of the coating or paint, and to determine whether any cleaning, surface preparation, or conditioning is necessary to ensure accurate results. In some cases, it may be necessary to remove the coating or paint to perform the inspection. Additionally, it is important to use a penetrant with a low viscosity that can easily conform to the surface of the part without leaving residue.

How do you handle penetrant testing of parts that have been welded?

Answer: Handling penetrant testing of parts that have been welded can be challenging due to the potential for additional defects or variations in the material. To handle this, it is important to have a well-defined inspection plan that includes the proper positioning and handling of the part, as well as the use of appropriate equipment and personnel.

Before the inspection, it's important to evaluate the condition of the weld and to determine whether any cleaning, surface preparation, or conditioning is necessary to ensure accurate results. It may also be necessary to use specialized inspection techniques such as radiography or ultrasonic testing to evaluate the quality of the weld. Additionally, it is important to document the weld history of the part, and to compare the inspection results with this history to ensure that the weld has not introduced new defects.

How do you handle penetrant testing of parts that have been plated?

Answer: Handling penetrant testing of parts that have been plated can be challenging due to the difficulty of maintaining proper contact between the penetrant and the surface of the part during the inspection. To handle this, it is important to have a well-defined inspection plan that includes the proper positioning and handling of the part, as well as the use of appropriate equipment and personnel.

Before the inspection, it's important to evaluate the condition of the plating and to determine whether any cleaning, surface preparation, or conditioning is necessary to ensure accurate results. In some cases, it may be necessary to remove the plating to perform the inspection. Additionally, it is important to use a penetrant with a low viscosity that can easily conform to the surface of the part without leaving residue.

How do you handle penetrant testing of parts that have been heat treated?

Answer: Handling penetrant testing of parts that have been heat treated can be challenging due to the potential for additional defects or variations in the material. To handle this, it is important to have a well-defined inspection plan that includes the proper positioning and handling of the part, as well as the use of appropriate equipment and personnel.

Before the inspection, it's important to evaluate the condition of the part and to determine whether any cleaning, surface preparation, or conditioning is necessary to ensure accurate results. It may also be necessary to use specialized inspection techniques such as radiography or ultrasonic testing to evaluate the quality of the heat treatment. Additionally, it is important to document the heat treatment history of the part, and to compare the inspection results with this history to ensure that the heat treatment has not introduced new defects.

How do you handle penetrant testing of parts that have been exposed to high levels of radiation?

Answer: Handling penetrant testing of parts that have been exposed to high levels of radiation can be challenging due to the potential for additional defects or variations in the material and the potential safety risks to the personnel performing the inspection. To handle this, it is important to have a well-defined inspection plan that includes the proper positioning and handling of the part, as well as the use of appropriate equipment and personnel.

Before the inspection, it's important to evaluate the level of radiation and to determine whether any cleaning, surface preparation, or conditioning is necessary to ensure accurate results.

It may also be necessary to use specialized inspection techniques such as radiography or ultrasonic testing to evaluate the quality of the part. Additionally, it is important to follow proper safety procedures, such as using personal protective equipment and limiting the amount of time spent near the part, to ensure the safety of the personnel involved.

How do you handle penetrant testing of parts that are large and heavy?

Answer: Handling penetrant testing of parts that are large and heavy can be challenging due to the difficulty of positioning and handling the part during the inspection. To handle this, it is important to have a well-defined inspection plan that includes the proper positioning and handling of the part, as well as the use of appropriate equipment and personnel. It may also be necessary to use specialized equipment such as cranes or hoists to move the part during the inspection process.

Additionally, it is important to have a sufficient number of trained personnel to assist with the handling and positioning of the part, and to use proper safety procedures to ensure the safety of the personnel involved.

How do you handle penetrant testing of parts that are small and delicate?

Answer: Handling penetrant testing of parts that are small and delicate can be challenging due to the difficulty of maintaining proper contact between the penetrant and the surface of the part during the inspection. To handle this, it is important to have a well-defined inspection plan that includes the proper positioning and handling of the part, as well as the use of appropriate equipment and personnel.

It may also be necessary to use specialized equipment such as microscope or magnifying lens to ensure precise inspection, and to use a low viscosity penetrant that can easily conform to the surface of the part without leaving residue. Additionally, it is important to use proper handling techniques and to have a sufficient number of trained personnel to assist with the handling and positioning of the part, and to use proper safety procedures to ensure the safety of the personnel involved.

How do you handle penetrant testing of parts that are designed to be non-magnetic?

Answer: Handling penetrant testing of parts that are designed to be non-magnetic can be challenging due to the need to ensure that the part remains non-magnetic during the inspection process. To handle this, it is important to have a well-defined inspection plan that includes the proper positioning and handling of the part, as well as the use of appropriate equipment and personnel.

It may also be necessary to use specialized inspection techniques such as eddy current testing or remote-field eddy current testing, which do not require the use of a magnetic field and therefore will not affect the non-magnetic properties of the part. Additionally, it is important to document the non-magnetic requirements of the part, and to compare the inspection results with these requirements to ensure that the part remains non-magnetic.

How do you handle penetrant testing of parts that are designed to be radioactive?

Answer: Handling penetrant testing of parts that are designed to be radioactive can be challenging due to the potential safety risks to the personnel performing the inspection. To handle this, it is important to have a well-defined inspection plan that includes the proper positioning and handling of the part, as well as the use of appropriate equipment and personnel.

Before the inspection, it's important to evaluate the level of radioactivity and to determine whether any cleaning, surface preparation, or conditioning is necessary to ensure accurate results. It may also be necessary to use specialized inspection techniques such as radiography or ultrasonic testing to evaluate the quality of the part. Additionally, it is important to follow proper safety procedures, such as using personal protective equipment and limiting the amount of time spent near the part, to ensure the safety of the personnel involved.

How do you evaluate the results of penetrant testing?

Answer: The results of penetrant testing are evaluated by comparing the indication produced on the surface of the part to the location, size, and shape of the actual defects. This evaluation is typically done by a trained and experienced inspector using a magnifying lens or microscope, and may also involve the use of specialized equipment such as a borescope or endoscope to access internal geometries of the part.

Additionally, it is important to document the results of the inspection, including any indications found and their locations, and to compare these results with the inspection plan and acceptance criteria to determine if any further action is necessary.

It is worth noting that the above-mentioned questions and answers are not exhaustive and may vary depending on the specific context and requirements of the testing. It is also important to note that while I am providing information on NDT penetrant testing, it is important to consult with a certified and qualified professional to ensure the safety and accuracy of the inspection process.

How do you ensure the reliability and accuracy of penetrant testing results?

Answer: To ensure the reliability and accuracy of penetrant testing results, it is important to follow proper testing procedures and protocols, as well as to use appropriate equipment and personnel. This includes selecting the appropriate type of penetrant and developer for the material and surface finish of the part, properly preparing the surface of the part before testing, and following proper cleaning and inspection procedures.

Additionally, it is important to have a well-trained and experienced inspector evaluate the results of the test, and to document and record all findings. To further increase the reliability and accuracy of the test, it is recommended to use multiple testing methods and to cross-reference the results. The inspector should also be familiar with the specific industry standards and regulations that apply to the part being tested and ensure compliance with those standards.

How do you ensure the safety of personnel during penetrant testing?

Answer: To ensure the safety of personnel during penetrant testing, it is important to follow proper safety procedures and protocols, and to use appropriate equipment and personnel. This includes using personal protective equipment such as gloves, goggles, and respirators, as well as following proper handling procedures for the penetrant and developer materials.

Additionally, it is important to ensure that the inspection area is well-ventilated and to limit the amount of time spent near the part during the inspection process. It is also important to properly train and educate the personnel on the safety procedures and risks associated with penetrant testing. If the part being tested is radioactive or exposed to high levels of radiation, it is important to follow strict safety protocols and to use specialized equipment to limit the exposure to the personnel.

How do you maintain the equipment and materials used in penetrant testing?

Answer: To ensure the reliability and accuracy of penetrant testing results, it is important to properly maintain the equipment and materials used in the testing process. This includes regular cleaning and calibration of the equipment, as well as regular replacement of consumable items such as penetrant and developer materials.

It is also important to store the materials in a controlled environment, as per the manufacturer's instructions, to maintain their effectiveness.

Additionally, it is important to keep detailed records of the maintenance, calibration and expiration dates of the equipment and materials to ensure they are being used within their recommended time frame.

How do you handle and dispose of the penetrant and developer materials after testing?

Answer: Handling and disposing of the penetrant and developer materials after testing is important to ensure the safety of personnel and the environment. The materials should be properly sealed and stored in a designated area, away from heat and direct sunlight. It is also important to follow proper disposal procedures as per the manufacturer's instructions and local regulations for hazardous materials.

This may include properly neutralizing the materials before disposal, or arranging for professional disposal services to handle the materials. It is also important to keep detailed records of the disposal process to ensure compliance with regulations.

How do you handle penetrant testing of parts that are designed to be used in extreme temperatures?

Answer: Handling penetrant testing of parts that are designed to be used in extreme temperatures can be challenging due to the potential for damage to the part or the penetrant and developer materials during the inspection process. To handle this, it is important to have a well-defined inspection plan that includes the proper positioning and handling of the part, as well as the use of appropriate equipment and personnel.

Before the inspection, it's important to evaluate the operating temperature range of the part and to determine whether any cleaning, surface preparation, or conditioning is necessary to ensure accurate results. Additionally, it is important to use penetrant and developer materials that are specifically designed to be used in extreme temperatures and to follow the manufacturer's recommended storage and handling procedures.

How do you handle penetrant testing of parts that are made of multiple materials?

Answer: Handling penetrant testing of parts that are made of multiple materials can be challenging due to the potential for variations in the surface finish and the potential for the penetrant and developer materials to interact differently with each material. To handle this, it is important to have a well-defined inspection plan that includes the proper positioning and handling of the part, as well as the use of appropriate equipment and personnel.

Before the inspection, it's important to identify the different materials used in the part and to determine the appropriate penetrant and developer materials that can be used on each surface. Additionally, it is important to use specialized inspection techniques such as radiography or ultrasonic testing to evaluate the quality of the part.

How do you handle penetrant testing of parts that are designed to be used in a liquid environment?

Answer: Handling penetrant testing of parts that are designed to be used in a liquid environment can be challenging due to the potential for damage to the part or the penetrant and developer materials during the inspection process. To handle this, it is important to have a well-defined inspection plan that includes the proper positioning and handling of the part, as well as the use of appropriate equipment and personnel.

Before the inspection, it's important to evaluate the type of liquid environment the part will be used in and to determine whether any cleaning, surface preparation, or conditioning is necessary to ensure accurate results. Additionally, it is important to use penetrant and developer materials that are specifically designed to be used in a liquid environment and to follow the manufacturer's recommended storage and handling procedures. It is also important to ensure that the inspection area is properly sealed and protected from any potential liquid exposure.

How do you handle penetrant testing of parts that have complex geometries?

Answer: Handling penetrant testing of parts that have complex geometries can be challenging due to the difficulty of maintaining proper contact between the penetrant and the surface of the part during the inspection. To handle this, it is important to have a well-defined inspection plan that includes the proper positioning and handling of the part, as well as the use of appropriate equipment and personnel.

Before the inspection, it's important to evaluate the complex geometries of the part and to determine whether any cleaning, surface preparation, or conditioning is necessary to ensure accurate results. Additionally, it is important to use specialized equipment such as borescopes or endoscopes to access internal geometries of the part, and to use a penetrant with a low viscosity that can easily conform to the surface of the part without leaving residue.

How do you handle penetrant testing of parts that are designed to be used in a high-pressure environment?

Answer: Handling penetrant testing of parts that are designed to be used in a high-pressure environment can be challenging due to the potential for damage to the part or the penetrant and developer materials during the inspection process. To handle this, it is important to have a well-defined inspection plan that includes the proper positioning and handling of the part, as well as the use of appropriate equipment and personnel.

Before the inspection, it's important to evaluate the high-pressure environment the part will be used in and to determine whether any cleaning, surface preparation, or conditioning is necessary to ensure accurate results. Additionally, it is important to use penetrant and developer materials that are specifically designed to be used in a high-pressure environment and to follow the manufacturer's recommended storage and handling procedures. It is also important to ensure that the inspection area is properly sealed and protected from any potential high-pressure exposure.

How do you handle penetrant testing of parts that have been previously repaired or modified?

Answer: Handling penetrant testing of parts that have been previously repaired or modified can be challenging due to the potential for additional defects or variations in the material. To handle this, it is important to have a well-defined inspection plan that includes the proper positioning and handling of the part, as well as the use of appropriate equipment and personnel.

Before the inspection, it's important to evaluate the repair or modification history of the part and to determine whether any cleaning, surface preparation, or conditioning is necessary to ensure accurate results. Additionally, it is important to use specialized inspection techniques such as radiography or ultrasonic testing to evaluate the quality of the repair or modification. It is also important to document the repair or modification history of the part and to compare the inspection results with this history to ensure that the repair or modification has not introduced new defects.

How do you handle penetrant testing of parts that are made of composite materials?

Answer: Handling penetrant testing of parts that are made of composite materials can be challenging due to the potential for variations in the material properties and the potential for the penetrant and developer materials to interact differently with the composite materials. To handle this, it is important to have a well-defined inspection plan that includes the proper positioning and handling of the part, as well as the use of appropriate equipment and personnel.

Before the inspection, it's important to identify the different composite materials used in the part and to determine the appropriate penetrant and developer materials that can be used on each surface. Additionally, it is important to use specialized inspection techniques such as ultrasonic testing or x-ray to evaluate the quality of the part. It is also important to consider the environmental factors that the part may be exposed to and to choose penetrant and developer materials accordingly.

How do you handle penetrant testing of parts that have been previously coated?

Answer: Handling penetrant testing of parts that have been previously coated can be challenging due to the potential for variations in the surface finish and the potential for the penetrant and developer materials to interact differently with the coating. To handle this, it is important to have a well-defined inspection plan that includes the proper positioning and handling of the part, as well as the use of appropriate equipment and personnel.

Before the inspection, it's important to evaluate the coating history of the part and to determine whether any cleaning, surface preparation, or conditioning is necessary to ensure accurate results.

Additionally, it is important to use specialized inspection techniques such as radiography or ultrasonic testing to evaluate the quality of the coating. It is also important to document the coating history of the part and to compare the inspection results with this history to ensure that the coating has not introduced new defects.

How do you handle penetrant testing of parts that have been previously painted?

Answer: Handling penetrant testing of parts that have been previously painted can be challenging due to the potential for variations in the surface finish and the potential for the penetrant and developer materials to interact differently with the paint. To handle this, it is important to have a well-defined inspection plan that includes the proper positioning and handling of the part, as well as the use of appropriate equipment and personnel. Before the inspection, it's important to evaluate the painting history of the part and to determine whether any cleaning, surface preparation, or conditioning is necessary to ensure accurate results.

Additionally, it is important to use specialized inspection techniques such as radiography or ultrasonic testing to evaluate the quality of the paint.

It is also important to document the painting history of the part, and to compare the inspection results with this history to ensure that the paint has not introduced new defects. It is also important to consider the type of paint used, as some types of paint may require special penetrants or developers, or may not be suitable for penetrant testing at all.

How do you handle penetrant testing of parts that have been previously heat-treated?

Answer: Handling penetrant testing of parts that have been previously heat-treated can be challenging due to the potential for variations in the surface finish and the potential for the penetrant and developer materials to interact differently with the heat-treated material. To handle this, it is important to have a well-defined inspection plan that includes the proper positioning and handling of the part, as well as the use of appropriate equipment and personnel.

Before the inspection, it's important to evaluate the heat-treatment history of the part and to determine whether any cleaning, surface preparation, or conditioning is necessary to ensure accurate results. Additionally, it is important to use specialized inspection techniques such as radiography or ultrasonic testing to evaluate the quality of the heat-treated material. It is also important to document the heat-treatment history of the part and to compare the inspection results with this history to ensure that the heat treatment has not introduced new defects.

It is also important to consider the type of heat treatment used, as some types of heat-treatment may require special penetrants or developers, or may not be suitable for penetrant testing at all.

How do you handle penetrant testing of parts that are designed to be used in a high-vibration environment?

Answer: Handling penetrant testing of parts that are designed to be used in a high-vibration environment can be challenging due to the potential for damage to the part or the penetrant and developer materials during the inspection process. To handle this, it is important to have a well-defined inspection plan that includes the proper positioning and handling of the part, as well as the use of appropriate equipment and personnel.

Before the inspection, it's important to evaluate the high-vibration environment the part will be used in and to determine whether any cleaning, surface preparation, or conditioning is necessary to ensure accurate results. Additionally, it is important to use penetrant and developer materials that are specifically designed to be used in a high-vibration environment and to follow the manufacturer's recommended storage and handling procedures. It is also important to ensure that the inspection area is properly secured and protected from any potential high-vibration exposure.

How do you handle penetrant testing of parts that are designed to be used in a high-stress environment?

Answer: Handling penetrant testing of parts that are designed to be used in a high-stress environment can be challenging due to the potential for damage to the part or the penetrant and developer materials during the inspection process. To handle this, it is important to have a well-defined inspection plan that includes the proper positioning and handling of the part, as well as the use of appropriate equipment and personnel.

Before the inspection, it's important to evaluate the high-stress environment the part will be used in and to determine whether any cleaning, surface preparation, or conditioning is necessary to ensure accurate results. Additionally, it is important to use penetrant and developer materials that are specifically designed to be used in a high-stress environment and to follow the manufacturer's recommended storage and handling procedures. It is also important to ensure that the inspection area is properly secured and protected from any potential high-stress exposure.

It is worth mentioning that the above list of questions is not exhaustive and new questions may arise depending on the specific context and requirements of the testing. Additionally, it is important to note that while I am providing information on NDT penetrant testing, it is important to consult with a certified and qualified professional to ensure the safety and accuracy of the inspection process.

How do you handle penetrant testing of parts that have been previously coated with a non-metallic coating?

Answer: Handling penetrant testing of parts that have been previously coated with a non-metallic coating can be challenging as the penetrant may not be able to penetrate through the coating and detect defects beneath it. To handle this, it is important to have a well-defined inspection plan that includes the proper positioning and handling of the part, as well as the use of appropriate equipment and personnel.

Before the inspection, it's important to evaluate the coating history of the part and to determine whether any cleaning, surface preparation, or conditioning is necessary to ensure accurate results. Additionally, it is important to use specialized inspection techniques such as radiography or ultrasonic testing to evaluate the quality of the coating and detect defects beneath it. It is also important to document the coating history of the part, and to compare the inspection results with this history to ensure that the coating has not introduced new defects.

How do you handle penetrant testing of parts that have been previously coated with a metallic coating?

Answer: Handling penetrant testing of parts that have been previously coated with a metallic coating can be challenging as the penetrant may interact differently with the coating and may not be able to penetrate through it to detect defects beneath it. To handle this, it is important to have a well-defined inspection plan that includes the proper positioning and handling of the part, as well as the use of appropriate equipment and personnel.

Before the inspection, it's important to evaluate the coating history of the part and to determine whether any cleaning, surface preparation, or conditioning is necessary to ensure accurate results. Additionally, it is important to use specialized inspection techniques such as radiography or ultrasonic testing to evaluate the quality of the coating and detect defects beneath it. It is also important to document the coating history of the part, and to compare the inspection results with this history to ensure that the coating has not introduced new defects.

How do you handle penetrant testing of parts that are designed to be used in a corrosive environment?

Answer: Handling penetrant testing of parts that are designed to be used in a corrosive environment can be challenging due to the potential for damage to the part or the penetrant and developer materials during the inspection process. To handle this, it is important to have a well-defined inspection plan that includes the proper positioning and handling of the part, as well as the use of appropriate equipment and personnel.

Before the inspection, it's important to evaluate the corrosive environment the part will be used in and to determine whether any cleaning, surface preparation, or conditioning is necessary to ensure accurate results. Additionally, it is important to use penetrant and developer materials that are specifically designed to be used in a corrosive environment and to follow the manufacturer's recommended storage and handling procedures. It is also important to ensure that the inspection area is properly sealed and protected from any potential corrosive exposure.

It is important to note that in some cases, the corrosive environment may not allow for penetrant testing, and an alternative inspection method should be considered.

How do you handle penetrant testing of parts that have been previously welded?

Answer: Handling penetrant testing of parts that have been previously welded can be challenging due to the potential for variations in the surface finish and the potential for the penetrant and developer materials to interact differently with the welded material. To handle this, it is important to have a well-defined inspection plan that includes the proper positioning and handling of the part, as well as the use of appropriate equipment and personnel.

Before the inspection, it's important to evaluate the welding history of the part and to determine whether any cleaning, surface preparation, or conditioning is necessary to ensure accurate results. Additionally, it is important to use specialized inspection techniques such as radiography or ultrasonic testing to evaluate the quality of the weld. It is also important to document the welding history of the part, and to compare the inspection results with this history to ensure that the welding has not introduced new defects.

It is also important to consider the type of welding used, as some types of welding may require special penetrants or developers, or may not be suitable for penetrant testing at all.

How do you handle penetrant testing of parts that are designed to be used in a high-temperature environment?

Answer: Handling penetrant testing of parts that are designed to be used in a high-temperature environment can be challenging due to the potential for damage to the part or the penetrant and developer materials during the inspection process. To handle this, it is important to have a well-defined inspection plan that includes the proper positioning and handling of the part, as well as the use of appropriate equipment and personnel. Before the inspection, it's important to evaluate the high-temperature environment the part will be used in and to determine whether any cleaning, surface preparation, or conditioning is necessary to ensure accurate results. Additionally, it is important to use penetrant and developer materials that are specifically designed to be used in a high-temperature environment and to follow the manufacturer's recommended storage and handling procedures.

It is also important to ensure that the inspection area is properly sealed and protected from any potential high-temperature exposure. It is important to note that in some cases, the high-temperature environment may not allow for penetrant testing, and an alternative inspection method should be considered.

How do you handle penetrant testing of parts that have been previously machined?

Answer: Handling penetrant testing of parts that have been previously machined can be challenging due to the potential for variations in the surface finish and the potential for the penetrant and developer materials to interact differently with the machined material. To handle this, it is important to have a well-defined inspection plan that includes the proper positioning and handling of the part, as well as the use of appropriate equipment and personnel. Before the inspection, it's important to evaluate the machining history of the part and to determine whether any cleaning, surface preparation, or conditioning is necessary to ensure accurate results. Additionally, it is important to use specialized inspection techniques such as radiography or ultrasonic testing to evaluate the quality of the machining. It is also important to document the machining history of the part, and to compare the inspection results with this history to ensure that the machining has not introduced new defects.

It is also important to consider the type of machining used, as some types of machining may require special penetrants or developers, or may not be suitable for penetrant testing at all.

How do you handle penetrant testing of parts that have been previously treated with a surface treatment?

Answer: Handling penetrant testing of parts that have been previously treated with a surface treatment can be challenging due to the potential for variations in the surface finish and the potential for the penetrant and developer materials to interact differently with the treated material. To handle this, it is important to have a well-defined inspection plan that includes the proper positioning and handling of the part, as well as the use of appropriate equipment and personnel.

Before the inspection, it's important to evaluate the surface treatment history of the part and to determine whether any cleaning, surface preparation, or conditioning is necessary to ensure accurate results. Additionally, it is important to use specialized inspection techniques such as radiography or ultrasonic testing to evaluate the quality of the surface treatment. It is also important to document the surface treatment history of the part, and to compare the inspection results with this history to ensure that the surface treatment has not introduced new defects.

It is also important to consider the type of surface treatment used, as some types of surface treatment may require special penetrants or developers, or may not be suitable for penetrant testing at all.

How do you ensure that the penetrant and developer materials are properly stored and handled?

Answer: To ensure that the penetrant and developer materials are properly stored and handled, it is important to follow the manufacturer's recommended storage and handling procedures. This may include guidelines on storage temperature, humidity, and light exposure, as well as instructions for mixing and preparing the materials for use.

It is also important to ensure that the materials are stored in their original containers and that the containers are properly labeled and kept in a safe and secure location. It is also important to keep the materials away from any potential sources of contamination such as dust, chemicals, or other materials. Additionally, it is important to regularly check the expiration dates of the materials and to dispose of any expired materials in accordance with local regulations and guidelines.

How do you ensure that the penetrant testing equipment is properly maintained and calibrated?

Answer: To ensure that the penetrant testing equipment is properly maintained and calibrated, it is important to follow the manufacturer's recommended maintenance and calibration procedures. This may include regular cleaning and inspection of the equipment, as well as performing regular calibration checks to ensure accurate results.

It is also important to keep accurate records of the maintenance and calibration activities, including the date, equipment used, and results of the checks. Additionally, it is important to address any issues or repairs that may arise with the equipment promptly to ensure that it is in good working order. It is also important to ensure that the equipment is operated and maintained by personnel who are properly trained and qualified to use it.

How do you ensure that the penetrant testing process is performed in accordance with industry standards and regulations?

Answer: To ensure that the penetrant testing process is performed in accordance with industry standards and regulations, it is important to follow the relevant industry standards and guidelines, such as those established by organizations such as ASTM, ASME, or ISO.

This may include guidelines on the selection and use of penetrant and developer materials, the preparation of the part for inspection, and the interpretation of the inspection results. It is also important to ensure that the inspection process is performed by personnel who are properly trained and qualified to perform penetrant testing. Additionally, it is important to keep accurate records of the inspection process, including the date, equipment used, personnel involved, and results of the inspection. It is also important to comply with any relevant safety and environmental regulations to ensure the safety of the personnel and the protection of the environment.

How do you ensure the penetrant testing process is performed with high-quality and consistency?

Answer: To ensure the penetrant testing process is performed with high-quality and consistency, it is important to have a well-defined inspection plan in place and to follow it consistently. This plan should include guidelines on the selection and use of penetrant and developer materials, the preparation of the part for inspection, and the interpretation of the inspection results.

It is also important to ensure that the inspection process is performed by personnel who are properly trained and qualified to perform penetrant testing. A quality control program should be implemented to ensure that the inspection process is performed consistently and that the results are accurate. Additionally, it is important to have a clear procedure for reviewing and documenting the inspection results, and to have a process in place to address any issues or nonconformities that may arise.

How do you handle penetrant testing of parts that have been previously painted?

Answer: Handling penetrant testing of parts that have been previously painted can be challenging as the penetrant may not be able to penetrate through the paint and detect defects beneath it. To handle this, it is important to have a well-defined inspection plan that includes the proper positioning and handling of the part, as well as the use of appropriate equipment and personnel.

Before the inspection, it's important to evaluate the painting history of the part and to determine whether any cleaning, surface preparation, or conditioning is necessary to ensure accurate results. In some cases, it may be necessary to remove the paint in order to perform penetrant testing, this can be done with a paint remover or by abrasive blasting. Additionally, it is important to use penetrant and developer materials that are specifically designed to be used on painted surfaces and to follow the manufacturer's recommended storage and handling procedures.

It is also important to ensure that the inspection area is properly sealed and protected from any potential exposure to dust, chemicals or other materials that could affect the accuracy of the results.

How do you handle penetrant testing of parts that have been previously coated with a protective coating?

Answer: Handling penetrant testing of parts that have been previously coated with a protective coating can be challenging as the penetrant may not be able to penetrate through the coating and detect defects beneath it. To handle this, it is important to have a well-defined inspection plan that includes the proper positioning and handling of the part, as well as the use of appropriate equipment and personnel.

Before the inspection, it's important to evaluate the coating history of the part and to determine whether any cleaning, surface preparation, or conditioning is necessary to ensure accurate results. In some cases, it may be necessary to remove the coating in order to perform penetrant testing, this can be done with a coating remover or by abrasive blasting. Additionally, it is important to use penetrant and developer materials that are specifically designed to be used on coated surfaces and to follow the manufacturer's recommended storage and handling procedures. It is also important to ensure that the inspection area is properly sealed and protected from any potential exposure to dust, chemicals or other materials that could affect the accuracy of the results.

How do you handle penetrant testing of parts that have been previously coated with a non-conductive coating?

Answer: Handling penetrant testing of parts that have been previously coated with a non-conductive coating can be challenging as the penetrant may not be able to penetrate through the coating and detect defects beneath it. To handle this, it is important to have a well-defined inspection plan that includes the proper positioning and handling of the part, as well as the use of appropriate equipment and personnel.

Before the inspection, it's important to evaluate the coating history of the part and to determine whether any cleaning, surface preparation, or conditioning is necessary to ensure accurate results. In some cases, it may be necessary to remove the coating in order to perform penetrant testing, this can be done with a coating remover or by abrasive blasting. Additionally, it is important to use penetrant and developer materials that are specifically designed to be used on non-conductive surfaces and to follow the manufacturer's recommended storage and handling procedures. It is also important to ensure that the inspection area is properly sealed and protected from any potential exposure to dust, chemicals or other materials that could affect the accuracy of the results.

How do you handle penetrant testing of parts that have been previously coated with a conductive coating?

Answer: Handling penetrant testing of parts that have been previously coated with a conductive coating can be challenging as the conductive coating may interfere with the flow of the penetrant and affect the accuracy of the inspection results. To handle this, it is important to have a well-defined inspection plan that includes the proper positioning and handling of the part, as well as the use of appropriate equipment and personnel. Before the inspection, it's important to evaluate the coating history of the part and to determine whether any cleaning, surface preparation, or conditioning is necessary to ensure accurate results. In some cases, it may be necessary to remove the coating in order to perform penetrant testing, this can be done with a coating remover or by abrasive blasting. Additionally, it is important to use penetrant and developer materials that are specifically designed to be used on conductive surfaces and to follow the manufacturer's recommended storage and handling procedures.

It is also important to ensure that the inspection area is properly sealed and protected from any potential exposure to dust, chemicals or other materials that could affect the accuracy of the results.

How do you handle penetrant testing of parts that have been previously treated with a chemical treatment?

Answer: Handling penetrant testing of parts that have been previously treated with a chemical treatment can be challenging as the chemical treatment may affect the surface of the part and the penetration of the penetrant. To handle this, it is important to have a well-defined inspection plan that includes the proper positioning and handling of the part, as well as the use of appropriate equipment and personnel.

Before the inspection, it's important to evaluate the chemical treatment history of the part and to determine whether any cleaning, surface preparation, or conditioning is necessary to ensure accurate results. In some cases, it may be necessary to neutralize or remove the chemical treatment in order to perform penetrant testing, this can be done with a neutralizing solution or by abrasive blasting.

Additionally, it is important to use penetrant and developer materials that are specifically designed to be used on chemically treated surfaces and to follow the manufacturer's recommended storage and handling procedures. It is also important to ensure that the inspection area is properly sealed and protected from any potential exposure to dust, chemicals or other materials that could affect the accuracy of the results.

How do you handle penetrant testing of parts that have been previously treated with a heat treatment?

Answer: Handling penetrant testing of parts that have been previously treated with a heat treatment can be challenging as the heat treatment may affect the surface of the part and the penetration of the penetrant. To handle this, it is important to have a well-defined inspection plan that includes the proper positioning and handling of the part, as well as the use of appropriate equipment and personnel.

Before the inspection, it's important to evaluate the heat treatment history of the part and to determine whether any cleaning, surface preparation, or conditioning is necessary to ensure accurate results. In some cases, it may be necessary to cool the part down to room temperature before performing penetrant testing. Additionally, it is important to use penetrant and developer materials that are specifically designed to be used on heat-treated surfaces and to follow the manufacturer's recommended storage and handling procedures.

It is also important to ensure that the inspection area is properly sealed and protected from any potential exposure to dust, chemicals or other materials that could affect the accuracy of the results.

How do you ensure that the penetrant testing process is performed with minimal impact on the environment?

Answer: To ensure that the penetrant testing process is performed with minimal impact on the environment, it is important to follow proper disposal and handling procedures for the penetrant and developer materials. This may include guidelines on the storage and disposal of any hazardous materials used in the inspection process.

Additionally, it is important to use environmentally friendly materials whenever possible, and to ensure that any cleaning or surface preparation methods used do not release harmful chemicals or pollutants into the environment. It is also important to comply with any relevant environmental regulations to ensure the safety of the personnel and the protection of the environment.

How do you handle penetrant testing of parts that are difficult to access or have complex geometries?

Answer: Handling penetrant testing of parts that are difficult to access or have complex geometries can be challenging as it may require special equipment or techniques to properly position and inspect the part. To handle this, it is important to have a well-defined inspection plan that includes the proper positioning and handling of the part, as well as the use of appropriate equipment and personnel.

Before the inspection, it's important to evaluate the geometry of the part and to determine whether any cleaning, surface preparation, or conditioning is necessary to ensure accurate results. Additionally, it is important to use specialized inspection techniques such as radiography or ultrasonic testing to evaluate the quality of the part. It is also important to use specialized equipment such as borescopes, or to use remote visual inspection techniques to access and inspect the part. It is also important to consider the size and weight of the part and to ensure that the equipment and personnel are capable of handling it safely.

How do you handle penetrant testing of parts that are irregularly shaped or have complex features?

Answer: Handling penetrant testing of parts that are irregularly shaped or have complex features can be challenging as it may require special equipment or techniques to properly position and inspect the part. To handle this, it is important to have a well-defined inspection plan that includes the proper positioning and handling of the part, as well as the use of appropriate equipment and personnel.

Before the inspection, it's important to evaluate the shape and features of the part and to determine whether any cleaning, surface preparation, or conditioning is necessary to ensure accurate results. Additionally, it is important to use specialized inspection techniques such as radiography or ultrasonic testing to evaluate the quality of the part. It is also important to use specialized equipment such as borescopes, or to use remote visual inspection techniques to access and inspect the part.

It is also important to consider the size and weight of the part and to ensure that the equipment and personnel are capable of handling it safely.

How do you handle penetrant testing of parts that are very large or heavy?

Answer: Handling penetrant testing of parts that are very large or heavy can be challenging as it may require special equipment or techniques to properly position and inspect the part. To handle this, it is important to have a well-defined inspection plan that includes the proper positioning and handling of the part, as well as the use of appropriate equipment and personnel. Before the inspection, it's important to evaluate the size and weight of the part and to determine whether any cleaning, surface preparation, or conditioning is necessary to ensure accurate results. Additionally, it is important to use specialized inspection techniques such as radiography or ultrasonic testing to evaluate the quality of the part. It is also important to use specialized equipment such as cranes, hoists or other heavy lifting equipment to move and position the part, as well as to ensure that the equipment and personnel are capable of handling it safely.

How do you handle penetrant testing of parts that have been previously repaired or modified?

Answer: Handling penetrant testing of parts that have been previously repaired or modified can be challenging as it may require special equipment or techniques to properly position and inspect the part. To handle this, it is important to have a well-defined inspection plan that includes the proper positioning and handling of the part, as well as the use of appropriate equipment and personnel.

Before the inspection, it's important to evaluate the repair or modification history of the part and to determine whether any cleaning, surface preparation, or conditioning is necessary to ensure accurate results. Additionally, it is important to use specialized inspection techniques such as radiography or ultrasonic testing to evaluate the quality of the part, particularly in areas where repairs or modifications were made. It is also important to consider the size and weight of the part and to ensure that the equipment and personnel are capable of handling it safely.

How do you handle penetrant testing of parts that are made of dissimilar materials?

Answer: Handling penetrant testing of parts that are made of dissimilar materials can be challenging as it may require special equipment or techniques to properly position and inspect the part. To handle this, it is important to have a well-defined inspection plan that includes the proper positioning and handling of the part, as well as the use of appropriate equipment and personnel.

Before the inspection, it's important to evaluate the materials of the part and to determine whether any cleaning, surface preparation, or conditioning is necessary to ensure accurate results. Additionally, it is important to use specialized inspection techniques such as radiography or ultrasonic testing to evaluate the quality of the part, particularly in areas where dissimilar materials are joined. It is also important to consider the size and weight of the part and to ensure that the equipment and personnel are capable of handling it safely.

How do you ensure the penetrant testing equipment is properly calibrated and maintained?

Answer: To ensure the penetrant testing equipment is properly calibrated and maintained, it is important to have a well-defined maintenance plan in place. This plan should include guidelines on the calibration and maintenance of the equipment, as well as the frequency of these procedures. It is also important to have a process in place to address any issues or nonconformities that may arise.

Additionally, it is important to have the equipment calibrated by a qualified technician or service provider, following the manufacturer's recommended procedures. Furthermore, it is important to ensure that the equipment is properly stored, handled and cleaned after each use, to ensure its longevity and accuracy. It is also recommended to have a log book or record of the equipment's maintenance history, to be able to track the equipment's performance and quickly identify any potential issues.

How do you ensure that the penetrant testing process is carried out in accordance with the relevant standards and specifications?

Answer: To ensure that the penetrant testing process is carried out in accordance with the relevant standards and specifications, it is important to have a well-defined inspection plan that includes the proper positioning and handling of the part, as well as the use of appropriate equipment and personnel.

Additionally, it is important to be familiar with the relevant industry standards and specifications such as ASTM E1417, ASME BPVC, ISO 3452 and NAS 410. This includes adhering to guidelines on the selection of penetrant and developer materials, as well as the proper application and removal of the materials. It is also important to have a process in place to address any issues or nonconformities that may arise. Furthermore, it is important to have a written procedure describing the penetrant testing process and to provide training to the personnel performing the inspection, to ensure that the process is carried out consistently and in accordance with the relevant standards and specifications.

How do you ensure that the penetrant testing results are accurate and reliable?

Answer: To ensure that the penetrant testing results are accurate and reliable, it is important to have a well-defined inspection plan that includes the proper positioning and handling of the part, as well as the use of appropriate equipment and personnel. Additionally, it is important to have a process in place to address any issues or nonconformities that may arise.

It is also important to ensure that the equipment is properly calibrated and maintained, following the manufacturer's recommended procedures. Furthermore, it is important to adhere to guidelines on the selection of penetrant and developer materials, as well as the proper application and removal of the materials. Also, it is important to have the inspection process carried out by trained and qualified personnel, who understand the process and are able to detect defects and interpret the results. Finally, it is important to have a process in place to verify the results and to review the inspection process regularly to ensure that it is consistent and reliable.

How do you ensure that the penetrant testing process is safe for both the personnel and the parts being inspected?

Answer: To ensure that the penetrant testing process is safe for both the personnel and the parts being inspected, it is important to have a well-defined inspection plan that includes the proper positioning and handling of the part, as well as the use of appropriate equipment and personnel. Additionally, it is important to have a process in place to address any issues or nonconformities that may arise.

It is also important to ensure that the equipment is properly calibrated and maintained, following the manufacturer's recommended procedures. Furthermore, it is important to adhere to guidelines on the selection of penetrant and developer materials, as well as the proper application and removal of the materials. It is also important to have the inspection process carried out by trained and qualified personnel, who understand the process and are able to detect defects and interpret the results. It is also important to comply with safety regulations and guidelines, such as OSHA regulations, and to provide the personnel with proper personal protective equipment and training.

How do you ensure that the penetrant testing process is repeatable and consistent?

Answer: To ensure that the penetrant testing process is repeatable and consistent, it is important to have a well-defined inspection plan that includes the proper positioning and handling of the part, as well as the use of appropriate equipment and personnel.

Additionally, it is important to have a process in place to address any issues or nonconformities that may arise. It is also important to ensure that the equipment is properly calibrated and maintained, following the manufacturer's recommended procedures. Furthermore, it is important to adhere to guidelines on the selection of penetrant and developer materials, as well as the proper application and removal of the materials. It is also important to have the inspection process carried out by trained and qualified personnel, who understand the process and are able to detect defects and interpret the results.

It is also important to have a written procedure describing the penetrant testing process and to provide training to the personnel performing the inspection, to ensure that the process is carried out consistently. Finally, it is important to have a process in place to review the inspection process regularly to ensure that it is consistent and reliable.

How do you ensure that the penetrant testing process is cost-effective?

Answer: To ensure that the penetrant testing process is cost-effective, it is important to have a well-defined inspection plan that includes the proper positioning and handling of the part, as well as the use of appropriate equipment and personnel. Additionally, it is important to have a process in place to address any issues or nonconformities that may arise.

It is also important to ensure that the equipment is properly calibrated and maintained, following the manufacturer's recommended procedures. Furthermore, it is important to adhere to guidelines on the selection of penetrant and developer materials, as well as the proper application and removal of the materials. It is also important to have the inspection process carried out by trained and qualified personnel, who understand the process and are able to detect defects and interpret the results. Also, it is important to consider the cost of the process, including the cost of the materials, equipment, and personnel, and to make sure that the process is carried out in the most efficient and cost-effective way possible.

How do you handle penetrant testing of parts that have been previously painted or coated?

Answer: Handling penetrant testing of parts that have been previously painted or coated can be challenging as the paint or coating may affect the surface of the part and the penetration of the penetrant. To handle this, it is important to have a well-defined inspection plan that includes the proper positioning and handling of the part, as well as the use of appropriate equipment and personnel.

Before the inspection, it's important to evaluate the paint or coating history of the part and to determine whether any cleaning, surface preparation, or conditioning is necessary to ensure accurate results. This may include removing the paint or coating to expose the surface of the part, or using specialized techniques such as surface etching to prepare the surface for inspection. Additionally, it is important to use penetrant and developer materials that are specifically designed to be used on painted or coated surfaces and to follow the manufacturer's recommended storage and handling procedures. It is also important to ensure that the inspection area is properly sealed and protected from any potential exposure to dust, chemicals or other materials that could affect the accuracy of the results.

How do you ensure that the penetrant testing process is performed in a timely manner without compromising the quality of the inspection?

Answer: To ensure that the penetrant testing process is performed in a timely manner without compromising the quality of the inspection, it is important to have a well-defined inspection plan that includes the proper positioning and handling of the part, as well as the use of appropriate equipment and personnel.

Additionally, it is important to have a process in place to address any issues or nonconformities that may arise. It is also important to ensure that the equipment is properly calibrated and maintained, following the manufacturer's recommended procedures. Furthermore, it is important to adhere to guidelines on the selection of penetrant and developer materials, as well as the proper application and removal of the materials. It is also important to have the inspection process carried out by trained and qualified personnel, who understand the process and are able to detect defects and interpret the results. Finally, it is important to have a process in place to review the inspection process regularly to ensure that it is consistent and reliable. To avoid delays, it is also important to have a proper schedule in place, and to have adequate resources such as personnel, equipment and materials to carry out the inspection process in a timely manner.

How do you handle penetrant testing of parts that have been previously heat treated?

Answer: Handling penetrant testing of parts that have been previously heat treated can be challenging as the heat treatment may affect the surface of the part and the penetration of the penetrant. To handle this, it is important to have a well-defined inspection plan that includes the proper positioning and handling of the part, as well as the use of appropriate equipment and personnel.

Before the inspection, it's important to evaluate the heat treatment history of the part and to determine whether any cleaning, surface preparation, or conditioning is necessary to ensure accurate results. This may include allowing the part to cool to room temperature before inspection, or using specialized techniques such as surface etching to prepare the surface for inspection. Additionally, it is important to use penetrant and developer materials that are specifically designed to be used on heat-treated surfaces and to follow the manufacturer's recommended storage and handling procedures. It is also important to ensure that the inspection area is properly sealed and protected from any potential exposure to dust, chemicals or other materials that could affect the accuracy of the results.

How do you handle penetrant testing of parts that have complex geometries?

Answer: Handling penetrant testing of parts that have complex geometries can be challenging as it may require special equipment or techniques to properly position and inspect the part. To handle this, it is important to have a well-defined inspection plan that includes the proper positioning and handling of the part, as well as the use of appropriate equipment and personnel.

Before the inspection, it's important to evaluate the geometry of the part and to determine whether any cleaning, surface preparation, or conditioning is necessary to ensure accurate results. Additionally, it is important to use specialized inspection techniques such as radiography or ultrasonic testing to evaluate the quality of the part, particularly in areas that are difficult to access or inspect. It is also important to use specialized equipment such as flexible or articulated arms, or robotic systems to reach and inspect complex geometries. Additionally, it is important to provide proper training to the personnel performing the inspection, to ensure that the process is carried out consistently and in accordance with the relevant standards and specifications.

how do you ensure that the penetrant testing process is environmentally friendly?

answer: to ensure that the penetrant testing process is environmentally friendly, it is important to use materials and equipment that are non-toxic, biodegradable, and easy to dispose of. additionally, it is important to have a process in place to address any issues or nonconformities that may arise.

It is also important to ensure that the equipment is properly calibrated and maintained, following the manufacturer's recommended procedures. Furthermore, it is important to adhere to guidelines on the selection of penetrant and developer materials, as well as the proper application and removal of the materials. It is also important to have the inspection process carried out by trained and qualified personnel, who understand the process and are able to detect defects and interpret the results. It is also important to comply with environmental regulations and guidelines, and to minimize the use of hazardous materials and chemicals, and to dispose of them properly. Finally, it is important to have a process in place to review the inspection process regularly to ensure that it is consistent and reliable, and to minimize the environmental impact.

CHAPTER XI

Chapter 11: Most important Interviews Questions and Answers

What is the purpose of a dye penetrant?

The purpose of a dye penetrant is to detect surface-breaking defects in a material, such as cracks, porosity, or other flaws. It is commonly used in the manufacturing and maintenance of components for industries such as aerospace, automotive, and engineering.

The dye penetrant process involves cleaning the surface of the material to be inspected and then applying a liquid penetrant, which is drawn into any surface defects by capillary action. After a set amount of time, the excess penetrant is removed from the surface, and a developer is applied to draw out the penetrant trapped in the defects. This creates a visible indication of the flaw, which can be analyzed and evaluated.

Dye penetrant inspection is a non-destructive testing method, which means it does not damage the component being inspected. It is a quick, cost-effective way to detect surface defects and ensure the quality and safety of a product.

How do you check dye penetrant testing?

The process of checking dye penetrant testing involves several steps, which are as follows:

- Inspection: The first step in checking dye penetrant testing is to visually inspect the surface of the component that has been inspected. This is to identify any visible indications of defects or flaws that may have been missed during the inspection process.
- Cleaning: The next step is to clean the surface of the component thoroughly to remove any residual penetrant or developer that may be present.
- Developer application: The developer is then applied to the surface of the component, which draws out any penetrant that may be present in any surface-breaking defects. The developer typically creates a visible indication of the defect.

- Interpretation: The final step in checking dye penetrant testing is to interpret the indications produced by the developer. This involves evaluating the size, shape, location, and orientation of the indications to determine if they are defects or false indications. The evaluation is typically done by a qualified inspector or NDT technician who has been trained in the interpretation of dye penetrant inspection results.

It is important to note that the checking of dye penetrant testing is critical to ensuring the accuracy and reliability of the inspection results. Any defects or flaws that are missed or misinterpreted can have serious consequences, including safety risks and costly repairs or replacements.

Is there a dye penetrant examination acceptance criteria stated in ASME Section VIII?

Yes, ASME Section V, Nondestructive Examination, contains requirements for dye penetrant examination, including acceptance criteria. The acceptance criteria for dye penetrant examination are defined in Article 6 of ASME Section V, which covers liquid penetrant examination.

ASME Section V requires that the acceptance criteria for dye penetrant examination be specified in the applicable code, standard, or specification that governs the construction or maintenance of the component being inspected. The acceptance criteria may vary depending on the type and size of the component, the material being inspected, and the service conditions for which the component is intended.

For example, ASME Section VIII, Division 1 contains requirements for pressure vessels, and it specifies the acceptance criteria for dye penetrant examination in Mandatory Appendix 8 of the code. The acceptance criteria in Appendix 8 depend on the type and size of the defect, and they are based on the maximum allowable defect size that is permitted by the code.

It is important to note that the acceptance criteria for dye penetrant examination may vary depending on the applicable code or standard, and it is the responsibility of the inspector or NDT technician to ensure that the acceptance criteria are applied correctly.

Can a dye penetrant test be used in overhead welding?

Yes, dye penetrant testing can be used to inspect overhead welds. However, it may require additional precautions and considerations compared to inspecting welds in other positions.

When performing dye penetrant testing on overhead welds, it is important to ensure that the surface of the weld is clean and free of any debris or contaminants that could interfere with the inspection. This may require the use of special cleaning methods or equipment, depending on the type of weld and the location of the inspection.

In addition, it is important to select the appropriate type of penetrant and developer for the specific application, as well as to follow the manufacturer's instructions for application and inspection. The inspector should also be trained and qualified in the use of dye penetrant testing and in the inspection of overhead welds.

Overall, while there may be some additional challenges and considerations involved in using dye penetrant testing for overhead welds, it is a viable and effective method for detecting surface defects and ensuring the quality and safety of the welded component.

Why is the visible penetrant in dye penetrant testing red? Is there a specific reason behind choosing this colour?

The visible penetrant used in dye penetrant testing is not always red. In fact, penetrants are available in a variety of colors, including red, green, yellow, and blue. The choice of color is typically based on a combination of factors, such as personal preference, visibility against the background color of the material being inspected, and any specific requirements or standards that may apply.

That being said, red is a commonly used color for visible penetrants in dye penetrant testing. One reason for this is that red is a highly visible color that stands out against many different backgrounds. This can make it easier for the inspector to detect any surface-breaking defects that may be present.

Another reason for using a red penetrant is that it can be easily seen under black light, which is used in the developer step of the inspection process. When the developer is applied to the surface, it draws out the penetrant that has seeped into any surface-breaking defects, creating a visible indication of the flaw. When illuminated with a black light, the red penetrant fluoresces, making it easier to see and interpret the indications.

Overall, the choice of color for visible penetrants in dye penetrant testing will depend on a variety of factors, and there is no one "correct" color to use. The most important consideration is to choose a penetrant that provides good visibility and meets any applicable standards or requirements.

Can you explain the step-by-step dye penetrant testing process?

Here are the step-by-step instructions for performing dye penetrant testing:

- Surface preparation: The surface of the component to be inspected is first cleaned and dried to remove any dirt, grease, oil, or other contaminants that could interfere with the inspection.
- Penetrant application: A liquid penetrant is then applied to the surface of the component and allowed to seep into any surface-breaking defects that may be present. The penetrant is typically left on the surface for a specified amount of time, known as the dwell time, to allow sufficient penetration.
- Penetrant removal: The excess penetrant is then removed from the surface using a cleaner or solvent. The cleaner is designed to remove the penetrant from the surface while leaving it in any defects that may be present.
- Developer application: A developer is then applied to the surface of the component. The developer draws out any remaining penetrant that is trapped in surface-breaking defects, creating a visible indication of the flaw. The developer typically consists of a powder or liquid that is applied to the surface and allowed to dry.
- Inspection: The surface of the component is then inspected visually or using a black light, depending on the type of penetrant and developer used. The inspector looks for any visible indications of defects, such as cracks, porosity, or other surface irregularities.
- Interpretation: The final step in the process is to interpret the indications produced by the inspection. The inspector must evaluate the size, shape, location, and orientation of the indications to determine if they are defects or false indications. The evaluation is typically done by a qualified inspector or NDT technician who has been trained in the interpretation of dye penetrant inspection results.

It is important to note that the exact process may vary depending on the type of component being inspected, the type of penetrant and developer used, and any specific requirements or standards that may apply. However, these steps provide a general overview of the dye penetrant testing process.

Who can perform a dye penetrant inspection?

Dye penetrant inspection should be performed by qualified and trained personnel who have been certified in accordance with applicable industry standards and procedures. In general, the certification process involves training in the principles and practices of nondestructive testing (NDT), as well as practical training in the specific methods and techniques used for dye penetrant inspection.

The certification process may be conducted by a variety of organizations, such as industry associations, regulatory agencies, or independent certification bodies. The specific requirements for certification may vary depending on the industry and the type of component being inspected, but they typically include a combination of classroom training, hands-on training, and practical experience.

In addition to certification, personnel performing dye penetrant inspection should also be familiar with the specific standards and procedures that apply to the inspection, such as those established by ASME, ASTM, or other regulatory bodies. They should also have a good understanding of the properties and characteristics of the materials being inspected, as well as any factors that may affect the accuracy and reliability of the inspection results.

Overall, dye penetrant inspection should only be performed by trained and certified personnel who have the knowledge, skills, and experience to conduct the inspection safely and effectively.

On which materials can DPI (Dye Penetrant Inspection) be used?

Dye penetrant inspection (DPI) can be used on a wide range of non-porous materials, including metals, plastics, ceramics, and composites. However, the suitability of the method depends on the surface properties of the material and the type of defect that is being sought.

In general, DPI is most effective on materials that have a smooth, non-porous surface. This allows the penetrant to seep into any surface-breaking defects that may be present, creating a visible indication of the flaw. Porous materials, such as cast iron or some types of composites, may not be suitable for DPI because the penetrant can be absorbed into the material, making it difficult to detect surface-breaking defects.

It is also important to consider the compatibility of the penetrant with the material being inspected. Some penetrants may not be compatible with certain materials, which can result in false indications or damage to the material.

Overall, DPI is a versatile inspection method that can be used on a wide range of materials. However, the suitability of the method for a particular application will depend on a variety of factors, including the type of material being inspected, the type of defect that is being sought, and the specific requirements of the inspection standard or specification.

How do I find the size of the defect in dye penetrant test NDT?

Determining the size of a defect detected during a dye penetrant test (DPT) requires careful evaluation of the indications produced by the inspection. The size of the indication is typically measured and recorded in accordance with the inspection standard or specification being used.

The following steps can be used to determine the size of a defect detected during a DPT:

- Inspect the indication: The inspector should carefully examine the indication to determine its size, shape, location, and orientation. The indication may appear as a red line or other visible mark on the surface of the part.
- Use a measuring device: Depending on the size of the indication, the inspector may use a ruler, calipers, or other measuring device to determine its size.
- Record the size: The size of the indication should be recorded in accordance with the inspection standard or specification being used. This may involve measuring the length, width, and depth of the indication, or using a reference chart to estimate the size based on the appearance of the indication.
- Evaluate the significance: Once the size of the indication has been determined, the inspector should evaluate its significance in accordance with the inspection standard or specification being used. This may involve comparing the size of the indication to allowable limits or acceptance criteria to determine if the part is acceptable or requires further evaluation or repair.

It is important to note that the determination of defect size during a DPT requires the expertise of a trained and qualified inspector or NDT technician. The inspector must have a good understanding of the inspection standard or specification being used, as well as the properties and characteristics of the materials being inspected, to accurately evaluate the significance of any detected indications.

Why is magnetic particle inspection done when we have the dye penetration method?

Magnetic particle inspection (MPI) and dye penetrant testing (DPT) are both non-destructive testing methods that are commonly used to detect surface-breaking defects in metals. While they share some similarities, they also have some important differences that make them suited for different types of applications.

Here are some reasons why magnetic particle inspection is used when we have the dye penetration method:

- Detecting Subsurface Defects: MPI can detect subsurface defects in ferromagnetic materials, which cannot be detected using DPT. Subsurface defects can be caused by manufacturing defects, corrosion, or fatigue, and can weaken the material over time, leading to failure. MPI can detect these subsurface defects before they cause catastrophic failure.
- Greater Sensitivity: MPI is generally more sensitive than DPT for detecting smaller defects, such as cracks or fissures, which may not be detectable by DPT.
- Speed of Inspection: MPI can be a faster inspection method than DPT, as it can cover larger areas quickly, and the inspection can be conducted in real-time, making it ideal for high volume production environments.
- Equipment Availability: MPI equipment is generally more widely available than DPT equipment, making it more convenient to use in some situations.

While both methods are useful for detecting surface-breaking defects, they each have their own strengths and limitations, and the choice of method will depend on the specific requirements of the application. In many cases, it may be appropriate to use both methods in combination to ensure a more comprehensive inspection.

Explain Dye penetrant testing?

Dye penetrant testing (DPT), also known as liquid penetrant testing (LPT), is a non-destructive testing method that is used to detect surface-breaking defects in non-porous materials such as metals, plastics, ceramics, and composites. It is a relatively simple and low-cost method that can be used to detect surface defects such as cracks, porosity, laps, and other flaws that can compromise the integrity of a part.

The DPT process typically involves the following steps:

- Surface Preparation: The part to be inspected is thoroughly cleaned and dried to remove any dirt, oil, or other contaminants that may interfere with the test.

- Application of Penetrant: A liquid penetrant is applied to the surface of the part, either by spraying, brushing, or immersion. The penetrant is allowed to seep into any surface-breaking defects that may be present, typically for a specified dwell time.
- Penetrant Dwell Time: The penetrant is allowed to dwell on the surface for a specific time, typically ranging from a few minutes to several hours, depending on the penetrant and the material being inspected.
- Removal of Excess Penetrant: After the dwell time has elapsed, any excess penetrant is removed from the surface of the part using a cleaning solvent or water.
- Application of Developer: A developer is then applied to the surface of the part, which draws the penetrant out of any surface-breaking defects and spreads it out over the surface, making it visible to the inspector.
- Inspection: The part is inspected under adequate lighting, and any indications of defects are identified and evaluated based on the inspection standard or specification being used.

Dye penetrant testing is a versatile and widely used method for detecting surface-breaking defects in non-porous materials. It is relatively simple and easy to perform, making it ideal for the routine inspection of parts and structures. However, it is important to note that the interpretation of indications requires the expertise of a trained and qualified inspector or NDT technician, who can accurately evaluate the significance of any detected indications.

Can we use polyester cloth for cleaning purposes (for cleaning penetrant) in dye penetrant testing?

No, polyester cloth should not be used for cleaning purposes in dye penetrant testing. Polyester is a synthetic material that can leave lint or other fibers on the surface being cleaned, which can interfere with the penetrant and lead to false indications. In addition, polyester cloth can also absorb and retain penetrant, which can lead to contamination and interfere with subsequent inspections.

Instead, it is recommended to use lint-free and oil-free materials such as cotton or cellulose wipes, or lint-free, low-nap synthetic wipes specifically designed for penetrant testing. These materials should be free from any contaminants and should be used in a manner that prevents the introduction of additional contaminants into the part or surface being inspected.

It is important to follow the cleaning and inspection procedures specified in the applicable standards or specifications to ensure that the cleaning materials used are appropriate and do not interfere with the inspection process or the accuracy of the results.

What are the benefits of using ultrasonic testing over other methods, like dye penetrant or magnetic particle testing (MT)?

Ultrasonic testing (UT) offers several advantages over other non-destructive testing methods such as dye penetrant testing (DPT) or magnetic particle testing (MT). Some of the benefits of using UT include:

- Depth of Detection: Ultrasonic testing is capable of detecting internal defects within a material, whereas DPT and MT are primarily used for surface inspections. UT can detect flaws at greater depths than other NDT methods, making it particularly useful for inspecting thick-walled materials or components.
- Greater Sensitivity: Ultrasonic testing can detect very small flaws, including cracks and porosity, that may not be visible to the naked eye or are detectable by other NDT methods.
- Quantitative Results: Ultrasonic testing can provide accurate and quantitative measurements of the size, location, and depth of detected flaws. This makes it easier to determine the severity of defects and to plan appropriate repairs or maintenance.
- Versatility: Ultrasonic testing can be used to inspect a wide variety of materials, including metals, plastics, ceramics, and composites. It is also suitable for inspecting complex shapes and structures, making it a versatile and widely applicable NDT method.
- Non-Intrusive: Ultrasonic testing is a non-destructive testing method that does not damage the part being inspected. It is also relatively quick and easy to perform, making it a cost-effective option for routine inspections and maintenance.

Overall, ultrasonic testing offers several advantages over other NDT methods, particularly for detecting internal defects in thick-walled materials or components. It is a reliable and widely used method that provides accurate and quantitative results, making it an essential tool for many industries and applications.

How does a penetrant test has been done on Cessna 172R aircraft? And why?

The penetrant test, also known as dye penetrant inspection, is a non-destructive testing method that is commonly used in the aviation industry to detect surface cracks, porosity, and other defects in aircraft components. The Cessna 172R aircraft is a popular single-engine training and general aviation aircraft that is used around the world.

During the maintenance of the Cessna 172R aircraft, the penetrant test is typically used to inspect critical components such as the engine crankshaft, propeller shaft, landing gear, and wing attach points. The inspection involves the following steps:

- The component is cleaned thoroughly using a solvent or other cleaning agent to remove any dirt, grease, or other contaminants that may interfere with the inspection.
- A penetrant solution is applied to the surface of the component. The solution is allowed to penetrate any surface cracks or defects for a specified amount of time.
- The excess penetrant is then wiped off, and a developer is applied to the surface. The developer draws the penetrant out of any defects, making them visible as a bright red indication.
- The component is inspected under UV light or black light, which enhances the visibility of any indications.

The penetrant test is performed on the Cessna 172R aircraft as part of routine maintenance and inspections to ensure the safety and reliability of the aircraft. Any defects or indications found during the inspection may require further evaluation or repair to ensure the continued safe operation of the aircraft.

What NDT carried out in the Flow Line 36-inch pipeline?

Non-destructive testing (NDT) is commonly used in the oil and gas industry to inspect pipelines for defects or anomalies that may affect their structural integrity or performance. The specific NDT methods used in the inspection of a 36-inch pipeline may vary depending on the type of pipeline, its location, and the specific requirements of the inspection. However, some of the most common NDT methods used in pipeline inspection include:

- Ultrasonic Testing (UT): Ultrasonic testing is used to detect flaws and defects in the pipeline wall, including cracks, corrosion, and other defects that may affect its structural integrity.

- Magnetic Particle Inspection (MPI): Magnetic particle inspection is used to detect surface and near-surface defects in the pipeline, such as cracks and other anomalies that may affect the pipeline's performance.
- Radiographic Testing (RT): Radiographic testing involves the use of X-rays or gamma rays to detect defects and anomalies in the pipeline wall, including corrosion, cracks, and other defects that may affect its structural integrity.
- Eddy Current Testing (ECT): Eddy current testing is used to detect surface and near-surface defects in the pipeline, including cracks and other anomalies that may affect its performance.

The specific NDT methods used in the inspection of the Flow Line 36-inch pipeline may vary depending on the specific requirements of the inspection and the type of pipeline involved. However, these methods are commonly used in pipeline inspection and can provide valuable information about the condition of the pipeline and any defects that may need to be addressed to ensure its safe operation.

What are the non-destructive ways to check the quality of welding?

There are several non-destructive ways to check the quality of welding, including:

- Visual Inspection: Visual inspection is the most common non-destructive method used to check the quality of welding. It involves a thorough visual examination of the weld to check for any visible defects such as cracks, porosity, or incomplete fusion.
- Radiographic Testing (RT): Radiographic testing involves the use of X-rays or gamma rays to create an image of the weld. This method can detect internal defects such as porosity, inclusions, and lack of fusion.
- Ultrasonic Testing (UT): Ultrasonic testing involves the use of high-frequency sound waves to detect defects within the weld. This method can detect internal defects such as cracks, inclusions, and incomplete fusion.
- Magnetic Particle Inspection (MPI): Magnetic particle inspection is used to detect surface-breaking defects such as cracks in the weld. It involves the application of a magnetic field to the weld, followed by the application of iron particles that will be attracted to any magnetic flux leakage caused by a defect.

- Dye Penetrant Inspection (DPI): Dye penetrant inspection involves the application of a liquid dye to the surface of the weld. The dye penetrates any surface defects, and excess dye is wiped off before the application of a developer. The developer draws the dye out of any defects, making them visible.

By using these non-destructive methods, welds can be checked for defects and other quality issues without causing any damage to the welded joint. These methods allow for the early detection of defects, which can help prevent equipment failure and ensure the safety and reliability of the welded structure.

What are the types of non-destructive tests used for detecting welding defects?

There are several types of non-destructive tests (NDT) that can be used for detecting welding defects. Some of the most common NDT methods used for welding inspection include:

- Visual Inspection: This is the most basic type of NDT and involves a visual examination of the welded joint to check for surface defects such as cracks, porosity, incomplete fusion, or other imperfections.
- Radiographic Testing (RT): This method uses X-rays or gamma rays to create an image of the welded joint. The image can show internal defects such as cracks, porosity, inclusions, and other types of imperfections.
- Ultrasonic Testing (UT): This method uses high-frequency sound waves to detect internal defects within the welded joint, such as cracks, incomplete fusion, or porosity.
- Magnetic Particle Inspection (MPI): This method involves the application of a magnetic field to the welded joint. Iron particles are then applied to the surface, which is attracted to any magnetic flux leakage caused by a defect such as a crack.
- Liquid Penetrant Inspection (LPI): This method involves applying a liquid dye to the surface of the welded joint. The dye penetrates into any surface-breaking defects and excess dye is then removed before the application of a developer. The developer draws the dye out of the defects, making them visible.
- Eddy Current Testing (ECT): This method uses an alternating current passed through a probe, which creates an electromagnetic field around the welded joint. Any changes in the electromagnetic field caused by a

defect can be detected and analyzed.

These NDT methods are used to ensure that welding defects are detected early, allowing for corrective action to be taken before they result in equipment failure or other problems. By using these methods, welded joints can be inspected for defects and other quality issues without causing any damage to the welded structure.

What are the types of non-destructive tests used for detecting cracks and give us the properties of the material?

There are several types of non-destructive tests (NDT) that can be used for detecting cracks and determining the properties of the material. Some of the most common NDT methods used for crack detection and material property assessment include:

- Ultrasonic Testing (UT): This method uses high-frequency sound waves to detect cracks and other defects within the material. It can also be used to determine material properties such as the thickness and composition.
- Radiographic Testing (RT): This method uses X-rays or gamma rays to create an image of the material, which can show internal cracks and other defects. It can also be used to determine the material properties such as density and thickness.
- Eddy Current Testing (ECT): This method uses electromagnetic waves to detect cracks and other defects near the surface of the material. It can also be used to determine material properties such as electrical conductivity.
- Magnetic Particle Inspection (MPI): This method involves the application of a magnetic field to the material. Iron particles are then applied to the surface, which are attracted to any magnetic flux leakage caused by a defect such as a crack.
- Penetrant Testing (PT): This method involves applying a liquid dye to the surface of the material. The dye penetrates into any surface-breaking defects such as cracks and excess dye is then removed before the application of a developer. The developer draws the dye out of the defects, making them visible.

These NDT methods can be used to detect cracks and other defects in materials and provide information about the material properties such as thickness, composition, and electrical conductivity. By using these methods, defects can be detected early, allowing for corrective action to be taken before they result in equipment failure or other problems.

What is NDT in welding?

NDT in welding stands for Non-Destructive Testing, which is a group of testing techniques used to evaluate the quality and integrity of welded components or structures without causing any damage to them. NDT is an important aspect of welding quality control and assurance, as it helps to ensure that the welded components or structures meet the required quality standards and safety regulations.

NDT in welding involves using various testing methods such as ultrasonic testing, radiography, magnetic particle inspection, and liquid penetrant testing, among others. These methods are used to detect welding defects such as cracks, porosity, incomplete fusion, lack of penetration, and other imperfections that may affect the quality and integrity of the weld.

NDT in welding is carried out at different stages of the welding process, including before welding, during welding, and after welding. It is important to carry out NDT in welding to detect defects early, allowing for corrective actions to be taken before they result in equipment failure or other problems. NDT in welding is also important in ensuring that the welded components or structures are safe to use and meet the required quality standards.

What are the acceptance criteria for liquid penetrant testing?

The acceptance criteria for liquid penetrant testing (LPT) may vary depending on the applicable industry standards, specific material or component requirements, and the level of sensitivity required for the testing. However, the general acceptance criteria for LPT are based on the evaluation of the size, shape, and location of the detected indications or defects.

In general, the acceptance criteria for LPT are based on the size and type of defect detected. The size of the defect is compared to a standard reference block or calibration defect, and if the size of the detected defect exceeds the specified limit, the component or material may be rejected. The type of defect is also important in determining the acceptability of the material or component. Some defects, such as cracks or porosity, are generally considered unacceptable and may result in rejection.

The acceptance criteria for LPT are also dependent on the sensitivity level of the test method. For example, higher sensitivity levels may be required for critical components or materials, while lower sensitivity levels may be acceptable for less critical applications.

It is important to note that the acceptance criteria for LPT may be different for different materials, components, or industries, and should be established based on the applicable standards and requirements.

Why is a darkened area required for a water-washable visible penetrant system in liquid penetrant testing?

In liquid penetrant testing, a darkened area is required for water washable visible penetrant systems to enhance the visibility of the penetrant indications or defects.

Water-washable visible penetrant systems use a water-soluble dye as the penetrant, which is visible under white light. The penetrant is applied to the surface of the component being inspected and allowed to soak into any surface defects, cracks, or other imperfections. After a sufficient penetration time, the excess penetrant is removed from the surface, and a developer is applied to the surface to draw out any penetrant trapped in the defects.

To enhance the visibility of the penetrant indications, a darkened area is typically required for the inspection process. This is because the visible penetrant may be difficult to see under normal lighting conditions due to its relatively low contrast against the background. By reducing the ambient light in the inspection area, the contrast between the penetrant indications and the background is increased, making the indications easier to see.

The darkened area can be achieved by using curtains, enclosures, or other light-blocking materials to create a controlled environment with minimal ambient light. This can help to improve the sensitivity and accuracy of the inspection process by allowing the inspector to more easily identify and evaluate any penetrant indications or defects.

What is the importance of non-destructive testing?

Non-destructive testing (NDT) is important for several reasons:

- Safety: NDT helps to ensure the safety of people, equipment, and infrastructure. It allows defects and flaws to be detected and evaluated without causing damage to the component or structure being inspected. This enables the identification of potential hazards, such as cracks or corrosion, before they can lead to catastrophic failures.
- Reliability: NDT can improve the reliability and performance of materials and structures. By detecting and monitoring defects, it enables proactive maintenance and repair, which can extend the useful life of

components and structures.

- Quality control: NDT is an essential tool for quality control in the manufacturing and construction industries. It enables the detection and removal of defects and inconsistencies in materials and products before they are put into service, which can help to reduce rework and warranty costs.
- Cost savings: NDT can help to save costs by reducing the need for expensive and time-consuming destructive testing methods. It can also reduce the risk of failures and accidents, which can lead to significant financial and reputational losses.
- Regulatory compliance: Many industries are subject to regulatory requirements for NDT. Compliance with these requirements helps to ensure the safety, reliability, and quality of products and services, and can help to prevent legal and financial penalties.

Overall, NDT is important because it enables the safe, reliable, and cost-effective operation of infrastructure, equipment, and products, while also ensuring compliance with regulatory requirements and industry standards.

What are destructive and non-destructive tests?

Destructive testing (DT) and non-destructive testing (NDT) are two categories of testing methods used to evaluate the properties, performance, and quality of materials and structures.

- Destructive testing involves the physical destruction or alteration of the test specimen or component to evaluate its properties. Examples of DT methods include tensile testing, compression testing, bending testing, impact testing, and fatigue testing. These methods provide accurate and detailed information about the properties of the material, but they typically require the destruction of the specimen, which can be expensive and time-consuming.
- Non-destructive testing, on the other hand, evaluates the properties of materials and structures without causing any damage or alteration to the component being tested. NDT methods include visual inspection, ultrasonic testing, radiographic testing, magnetic particle testing, and liquid penetrant testing, among others. These methods are used to detect and evaluate flaws, defects, and discontinuities in materials and structures without affecting their performance or structural integrity.

Both DT and NDT methods have their advantages and limitations, and the choice of method depends on the specific requirements of the application. Destructive testing is typically used when accurate and detailed information about the properties of the material is required, while non-destructive testing is used for the inspection and evaluation of materials and structures without causing any damage or alteration.

What is the disadvantage of non-destructive testing?

Non-destructive testing (NDT) has several advantages over traditional destructive testing, as it allows the inspection and evaluation of materials and components without causing any damage or impairment to the item being tested. However, there are also some disadvantages of non-destructive testing that should be taken into account. Here are a few:

- Limitations in detecting certain types of defects: Some NDT methods may not be effective in detecting certain types of defects or flaws, such as subsurface defects or defects located in areas that are difficult to access or inspect. In such cases, destructive testing may be necessary to accurately detect and evaluate the extent of the defect.
- Expertise and training requirements: Non-destructive testing requires skilled and trained personnel to operate the equipment and interpret the test results accurately. The lack of qualified personnel can lead to inaccurate or incomplete testing results, which can have serious consequences in safety-critical industries.
- Cost: Non-destructive testing equipment can be expensive to purchase and maintain, and may require specialized facilities and infrastructure to operate. As a result, the cost of NDT can be higher than traditional destructive testing methods.
- False positives and false negatives: Non-destructive testing may produce false positives (indicating a defect when there is none) or false negatives (failing to detect a defect that is present). These errors can lead to unnecessary repairs or missed defects, both of which can be costly and potentially dangerous in safety-critical applications.

Overall, non-destructive testing can be a useful tool in many applications, but it is important to consider its limitations and drawbacks when selecting a testing method.

What is the advantage of non-destructive testing?

Non-destructive testing (NDT) has several advantages over traditional destructive testing methods. Here are a few:

- Preservation of the tested object: The primary advantage of non-destructive testing is that it allows for the evaluation and inspection of materials and components without causing damage or impairment to the item being tested. This is particularly important for expensive, irreplaceable, or critical components, where damaging the item could be costly or even dangerous.
- Cost-effective: NDT can be a cost-effective alternative to destructive testing, as it eliminates the need for replacement or repair of the tested item after the evaluation process.
- Time-efficient: NDT methods are typically faster and require less preparation time than destructive testing, which often involves the disassembly or destruction of the tested item.
- Increased safety: Non-destructive testing methods can be performed on live systems, meaning that the testing can be carried out while the system is still in operation, reducing downtime and the need for system shutdowns.
- Wide range of applications: NDT can be used to test a variety of materials, from metals and ceramics to polymers and composites. There are many different NDT methods available, each with their own strengths and limitations, which can be tailored to suit the specific requirements of the application.

Overall, non-destructive testing can provide a valuable means of evaluating the quality and integrity of materials and components without causing damage or impairment, making it a useful tool in a wide range of applications.

Disclaimer:

The information and opinions contained in this book, "Penetrant Testing: Principles, Techniques, and Applications - Interview Q&A with Experts," are intended for educational and informational purposes only. The content of this book is not intended to be a substitute for professional advice, diagnosis, or treatment.

The authors, editors, and publishers of this book have made every effort to ensure that the information provided is accurate, up-to-date, and complete. However, they do not warrant or represent that the information contained in this book is free from errors or omissions, or that the information is suitable for any specific purpose.

The information contained in this book is subject to change without notice. The authors, editors, and publishers of this book do not accept any responsibility or liability for any losses or damages that may arise from the use of the information contained in this book, whether direct or indirect, consequential, or otherwise.

Readers are advised to seek professional advice before making any decisions or taking any action based on the information contained in this book. The authors, editors, and publishers of this book are not responsible for any errors, omissions, or inaccuracies in the information provided by the experts interviewed for this book.

The opinions expressed in this book are those of the authors and experts interviewed and do not necessarily reflect the views of the publishers or any other organization.

CHAPTER XII

Conclusion

As we come to the end of this book on penetrant testing, it is clear that this is a powerful technique for detecting surface-breaking defects in a wide range of materials and applications. Through a series of interviews with experts in the field, we have gained insight into the principles, techniques, and applications of penetrant testing, as well as the challenges and opportunities facing the industry today.

We have learned about the different types of penetrant testing, including visible, fluorescent, and dual-purpose methods, and the advantages and disadvantages of each. We have explored the various techniques used to apply penetrants, from manual spray and brush methods to automated systems, and we have discussed the importance of proper surface preparation and cleaning to ensure accurate results.

We have also examined the diverse applications of penetrant testing, from aerospace and automotive industries to medical devices and even art restoration. The experts we have spoken to have highlighted the critical role that penetrant testing plays in ensuring the safety and reliability of products, and the need for ongoing innovation and improvement in the field.

While penetrant testing is a powerful technique, it is not without its challenges. We have heard from experts about the need for continued education and training to ensure that technicians are equipped with the knowledge and skills necessary to carry out effective testing. We have also discussed the need for increased standardization and regulation in the industry to ensure consistency and quality across different applications.

Overall, this book has provided a comprehensive overview of penetrant testing, from its principles and techniques to its applications and challenges. It is clear that penetrant testing will continue to play a critical role in ensuring the safety and reliability of products across a range of industries, and that ongoing innovation and improvement will be necessary to meet the evolving needs of the field.

Thank You

www.ingramcontent.com/pod-product-compliance
Ingram Content Group UK Ltd.
Pitfield, Milton Keynes, MK11 3LW, UK
UKHW022018190726
13853UKWH00005B/1998

9 798889 866374